architecture &

construction

NATIONAL GEOGRAPHIC LEARNING

DELMAR
CENGAGE Learning®

Australia • Brazil • Japan • Korea • Mexico • Singapore • Spain • United Kingdom • United States

Architecture & Construction

Vice President, Career, Education,
and Training Editorial:
David Garza

Director of Learning Solutions:
Sandy Clark

Managing Editor: Larry Main

Associate Acquisitions Editor:
Katie Hall

Product Manager: Anne Prucha

Instructional Designer:
Nancy Petite

Associate Marketing Manager:
Jillian Borden

Production Director: Wendy Troeger

Production Manager: Mark Bernard

Senior Content Project Manager:
Glenn Castle

Director of Design: Bruce Bond

Editorial Assistant: Kaitlin Murphy

For product information and technology assistance, contact us at
Cengage Learning Customer & Sales Support, 1-800-354-9706.
For permission to use material from this text or product,
submit all requests online at **www.cengage.com/permissions.**
Further permissions questions can be e-mailed to
permissionrequest@cengage.com.

Library of Congress Control Number: 2012942893

ISBN-13: 978-1-133-96023-2
ISBN-10: 1-133-96023-5

Delmar
5 Maxwell Drive,
Clifton Park, NY 12065-2919
USA

Cengage Learning is a leading provider of customized learning solutions with office locations around the globe, including Singapore, the United Kingdom, Australia, Mexico, Brazil, and Japan. Locate your local office at **www.cengage.com/global**.

Cengage Learning products are represented in Canada by Nelson Education, Ltd.

To learn more about Delmar, visit **www.cengage.com/brookscole**
Purchase any of our products at your local college store or at our preferred online store **www.CengageBrain.com**.

Printed in the United States of America
1 2 3 4 5 6 7 16 15 14 13 12

Table *of* Contents

About the Series

Cengage Learning and National Geographic Learning are proud to present the *National Geographic Learning Reader Series*. This groundbreaking series is brought to you through an exclusive partnership with the National Geographic Society, an organization that represents a tradition of amazing stories, exceptional research, first-hand accounts of exploration, rich content, and authentic materials.

The series brings learning to life by featuring compelling images, media, and text from National Geographic. Through this engaging content, students develop a clearer understanding of the world around them. Published in a variety of subject areas, the *National Geographic Learning Reader Series* connects key topics in each discipline to authentic examples and can be used in conjunction with most standard texts or online materials available for your courses.

How the reader works

Each article is focused on one topic relevant to the discipline. The introduction provides context to orient students and focus questions that suggest ideas to think about while reading the selection. Rich photography, compelling images, and pertinent maps are amply used to further enhance understanding of the selections. The chapter culminating section includes discussion questions to stimulate both in-class discussion and out-of-class work.

A premium eBook will accompany each reader and will provide access to the text online with a media library that may include images, videos and other premium content specific to each individual discipline.

National Geographic Learning Readers are currently available in a variety of course areas, including Archeology, Architecture & Construction, Biological Anthropology, Biology, Earth Science, English Composition, Environmental Science, Geography, Geology, Meteorology, Oceanography, and Sustainability.

Few organizations present this world, its people, places, and precious resources in a more compelling way than National Geographic. Through this reader series we honor the mission and tradition of National Geographic Society: to inspire people to care about the planet.

Preface

Modern technology and the green movement have changed t
way we think about architecture and construction. Our desi
to preserve the world and its resources has transformed new constru
tion as we look for ways to minimize our carbon footprint. The artic
in this reader are a testament to the work being done in the field
architecture and construction to become greener, to increase efficien
to meet the demands of our busy lives, and to preserve the archite
ture created by our ancestors. Readers are asked to explore the wa
each individual can make a difference in the decisions that they ma
daily in their home or in designing a home, building, or even a tunn
The articles and images presented herein remind us that we can ha
a positive impact on our changing world and leave a legacy of sustai
able, beautiful, and innovative architecture.

To begin, readers are introduced to some of the methods used
safeguard homes from natural disaster. "The Big Idea: Safe House
explores the way the construction of a house can impact its ability
stand up to an earthquake while "How to Help: Don't Fan the Flame
examines how a house can be constructed more safely when in
wildfire zone. Next, the discussion turns to large construction projec
like the Gotthard Base Tunnel in the Swiss Alps in "Tunnel Vision" ar
the Beijing architecture built specifically for the Olympics in "The Ne
Great Walls." Green issues are addressed in "Saving Energy: It Starts
Home," which asks the question: Will the efforts of individuals to sa
money and make energy-efficient decisions at home actually ma
any impact on the emissions of carbon dioxide? Then the followi
two articles look at the science behind preserving our past and o
future. "Next: Simulating Wildfires" highlights the research effor
taken to make a home less vulnerable to a wildfire. "Technology: Fu
Tilt" presents the science behind leaning towers and the preservatio
methods now being implemented to keep towers standing. Final
to round out the reader's dive into architecture and constructio
"Village Green" discusses how communities worldwide are addressir
the increasing need for energy, both as a community and from a
individual perspective.

This National Geographic reader brings together a diverse group
investigations into technology and highlights how recent research
revolutionizing our practices in architecture and construction. Ea
unit challenges the student to think beyond the content in the arti
and to consider how these advances impact them closer to hon
An Anticipation Guide introduces each unit, which then conclud
with a Writing Assignment, Career Investigation, and Team Buildi
Activity. Instructors can choose additional activities that are availab

n the Instructor Companion Website. The Writing Assignment asks
or further thought about the article and the impact that it has locally.
he Career Investigation highlights careers relevant to the individual
rticles and allows readers the chance to further explore the educational
equirements, essential skills needed, job descriptions and pay scales
f a variety of occupations. The Team Building Activity asks four or
ve students to form a team and research a topic further. These up-to-
ate and relevant *National Geographic* articles help students gain a
erspective on how the choices they make daily can impact their local
ommunity, environment, and even the world within the Architecture
nd Construction field.

ANTICIPATION GUIDE FOR:
THE BIG IDEA: SAFE HOUSES

Purpose: To identify what you already know about building construction, to direct and personalize your reading, and to provide a record of what new information you learned.

———————◆———————

Before you read about earthquake engineering in "The Big Idea: Safe Houses," examine each statement below and indicate whether you agree or disagree. Be prepared to discuss your reactions to the statements in groups.

- Earthquake engineering applies only to buildings that have two or more floors.
- Earthquake engineering is only necessary in highly populated urban areas.
- The death toll for earthquakes is higher today than it was in the past 100–150 years.
- An earthquake with a higher magnitude will result in a higher death toll.
- The deaths that occur during an earthquake are a result of the ground shaking.
- Undeveloped countries/regions have fewer high-rise buildings, and therefore there is no need for earthquake engineering.
- Undeveloped countries/regions cannot afford to apply earthquake engineering to their existing and new buildings.
- Earthquake engineering involves sophisticated equipment and techniques that are unavailable in undeveloped countries/regions.
- Nearly every building can be made earthquake resistant using locally available materials.
- Earthquake engineering a building will prevent structural damage to that building during an earthquake.

THE BIG IDEA: SAFE HOUSES

By Chris Carroll

THE EARTHQUAKE IN HAITI WAS A REMINDER:

BILLIONS OF PEOPLE LIVE IN HOUSES THAT CAN'T STAND SHAKING.

YET SAFER ONES CAN BE BUILT CHEAPLY—USING STRAW, ADOBE, OLD TIRES—BY APPLYING A FEW GENERAL PRINCIPLES.

In Los Angeles, Tokyo, and other rich cities in fault zones, the added expense of making buildings earthquake resistant has become a fact of life. Concrete walls are reinforced with steel, for instance, and a few buildings even rest on elaborate shock absorbers. Strict building codes were credited with saving thousands of lives when a magnitude 8.8 quake hit Chile in late February. But in less developed countries like Haiti, where a powerful quake in January killed some 222,500 people and left more than a million homeless, conventional earthquake engineering is often unaffordable. "The devastation in Haiti wouldn't happen in a developed country," says engineer Marcial Blondet of the Catholic University of Peru, in Lima. Yet it needn't happen anywhere. Cheap solutions exist.

Blondet has been working on ideas since 1970, when an earthquake in Peru killed 70,000 or more, many of whom died when their houses crumbled around them. Heavy, brittle walls of traditional adobe—cheap, sun-dried brick—cracked instantly when the

> **You rebuild your house, but you don't bury anyone.**

ground started bucking. Subsequent shakes brought roofs thundering down. Blondet's research team has found that existing adobe walls can be reinforced with a strong plastic mesh installed under plaster; in a quake, those walls crack but don't collapse, allowing occupants to escape. "You rebuild your house, but you don't bury anyone," Blondet says. Plastic mesh could also work as a reinforcement for concrete walls in Haiti and elsewhere.

Other engineers are working on methods that use local materials. Researchers in India have successfully tested a concrete house reinforced with bamboo. A model house for Indonesia rests on ground-motion dampers designed by John van de Lindt of Colorado State University: old tires filled with bags of sand. Such a house might be only a third as strong as one built on more sophisticated shock absorbers, (Continued on page 8)

Adapted from "The Big Idea: Safe Houses" by Chris Carroll: National Geographic Magazine, June 2010.

	Pakistan	**Haiti**
MOST DESTRUCTIVE QUAKE	October 8, 2005	January 12, 2010
LOCATION	Northern Pakistan/Kashmir	Port-au-Prince area
MAGNITUDE	7.6	7.0
FATALITIES	75,000	222,500

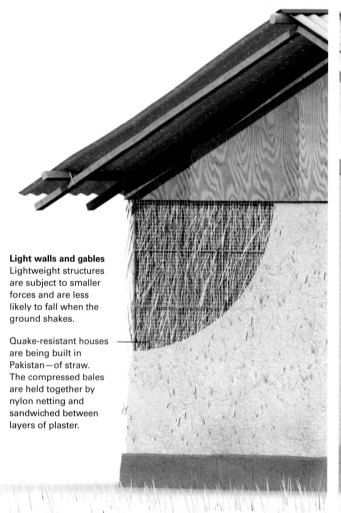

Light roofs
In Haiti heavy concrete roofs collapsed on many homes; in general, metal roofs on wooden trusses are more resilient.

Light walls and gables
Lightweight structures are subject to smaller forces and are less likely to fall when the ground shakes.

Quake-resistant houses are being built in Pakistan—of straw. The compressed bales are held together by nylon netting and sandwiched between layers of plaster.

Small windows
Small, regularly spaced openings create fewer weak spots in walls. But the bigger problem in Haiti was that walls were not properly reinforced.

Peru

May 31, 1970
Chimbote
7.9
70,000

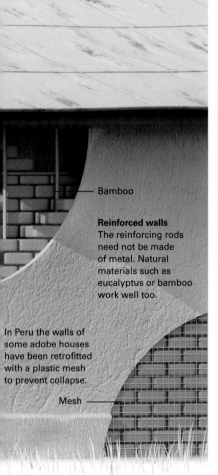

Bamboo

Reinforced walls
The reinforcing rods need not be made of metal. Natural materials such as eucalyptus or bamboo work well too.

In Peru the walls of some adobe houses have been retrofitted with a plastic mesh to prevent collapse.

Mesh

Indonesia

December 26, 2004
Sumatra
9.1
227,900 (includes global tsunami deaths)

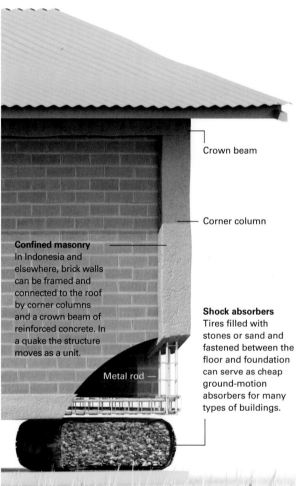

Crown beam

Corner column

Confined masonry
In Indonesia and elsewhere, brick walls can be framed and connected to the roof by corner columns and a crown beam of reinforced concrete. In a quake the structure moves as a unit.

Metal rod

Shock absorbers
Tires filled with stones or sand and fastened between the floor and foundation can serve as cheap ground-motion absorbers for many types of buildings.

ART: BRYAN CHRISTIE
SOURCES: GERNOT MINKE, UNIVERSITY OF KASSEL; ELIZABETH A. HAUSLER, BUILD CHANGE; ANNA LANG, UNIVERSITY OF CALIFORNIA, SAN DIEGO; MARCIAL BLONDET AND ÁLVARO RUBIÑOS, CATHOLIC UNIVERSITY OF PERU; PIERRE PAUL FOUCHÉ, UNIVERSITY AT BUFFALO; USGS

HAZARD ZONES *Many of the zones where the most intense earthquakes are likely to occur lie in less developed countries. It's not the violent shaking of the ground itself that claims the most victims, but the collapse of poorly constructed buildings.*

(Continued from page 5) but it would also cost much less—and so be more likely to get built in Indonesia. "As an engineer you ask, what level of safety do I need?" Van de Lindt says. "Then you look at what's actually available and find the solution somewhere in between."

In northern Pakistan, straw is available. Traditional houses are built of stone and mud, but straw is far more resilient, says California engineer Darcey Donovan, and warmer in winter to boot. Donovan and her colleagues started building straw-bale houses in Pakistan after the 2005 earthquake; so far they have completed 17.

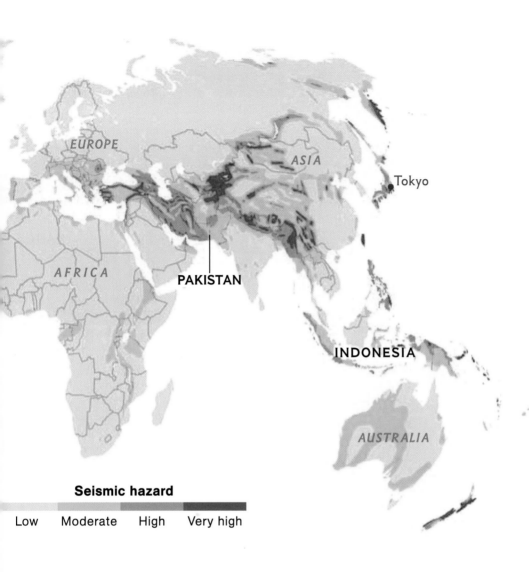

EUROPE

ASIA

Tokyo

AFRICA

PAKISTAN

INDONESIA

AUSTRALIA

Seismic hazard

Low   Moderate   High   Very high

The same stark contrast prevails in other fault zones: encouraging ideas, discouraging progress. Even cheap ideas aren't always cheap enough. Since 2007 some 2,500 houses in Peru have been strengthened with plastic mesh or other reinforcements, with another 700 scheduled for this year. That leaves millions of houses and billions of dollars to go in Peru alone, to say nothing of other countries. "There are many millions of houses around the world," Blondet says, "that will collapse in the next earthquake."

Writing Assignment

After reading the article "The Big Idea: Safe Houses" and conducting your own research, write a brief paper discussing design and construction practices that may be implemented to reduce the loss of life and property in earthquake prone regions.

In addition to your own ideas and thoughts, please discuss:

- Earthquake resistant strategies that have been implemented in earthquake prone regions other than Pakistan, Haiti, Peru, and Indonesia (you can compare construction techniques to Pakistan, Haiti, Peru, and Indonesia).

- The estimated cost of earthquake resistant strategies. Who would fund the program: homeowners or the governments?

Listed below are a few links to help you get started.

Links:

Indian Standards on Earthquake Engineering
http://www.bis.org.in/other/quake.htm

Housing Construction in Earthquake Prone Places
http://vps147.advomatic.com/sites/default/files/EJEM_Housing ConstructionInEarthquakePronePlaces.pdf

Izmit, Turkey Earthquake 1999
http://earthquake.usgs.gov/research/geology/turkey/index.php

Affordable Solution for Earthquake Resistant Building Construction in Haiti

http://www.mtnforum.org/sites/default/files/pub/affordable_earthequake-resistant.pdf

Return from the earthquake zone in Haiti
http://www2.gi.alaska.edu/ScienceForum/ASF20/2002.html

Earthquake-Resistant Construction of Adobe Buildings: A Tutorial
http://www.world-housing.net/uploads/WHETutorial_Adobe_English.pdf

Construction Guide on Strengthening Existing Houses in Hawaii Against Hurricanes and Earthquakes
http://www.seaoh.org/constructionguide.htm

Bamboo: Pitt team building school in India from the material
http://www.post-gazette.com/pg/09207/986211-437.stm

Career Investigation

Several recent high-magnitude earthquakes have resulted in large scale destruction and high death tolls. While it is impossible to predict exactly when an earthquake will strike, much can be done to mitigate the disastrous consequences.

Listed here are several possible careers to investigate. You may also be able to find a career path not listed. Choose three possible career paths and investigate what you would need to do to be ready to fill one of those positions.

Among the items to look for:

- Education—college? What would you major in? Should you have a minor?

- Working conditions—What is the day to day job like? Will you have to be out in the field all day, or is it a desk job? Is the work strenuous? Are you working on a drilling rig at sea, away from your family for 3–6 months at a time? Is the job hazardous?

- Pay scale—What is the average "starting pay" for the position you are seeking? Don't be fooled by looking at "average pay", which may include those who have been working for 20+ years.

- Is the job located in the states, or is there a possibility to travel to other parts of the world?

- Are there any other special requirements?

Architect
Architectural Technician or Technologist
Bricklayer
Building Surveyor
Building Technician
Carpenter or Joiner
Civil Engineer
Civil Engineering Technician
Construction Contracts Manager
Construction Manager
Electrician
Electricity Distribution Worker
Estimator
Geoscientist
Geotechnician
Heating and Ventilation Engineer

Insulation Installer
Land Surveyor
Painter and Decorator
Planning and Development Surveyor
Plasterer
Plumber
Project Manager
Quarry Engineer
Quarry Operative
Roofer
Rural Surveyor
Stonemason
Structural Engineer
Thatcher
Thermal Insulation Engineer
Tiler
Town Planner
Welder
Window Fitter
Wood Machinist

Team Building Activity

You will be assigned to a team of four and will select a country from the choices provided by the instructor and design a scale model structure using common building materials for the assigned country that will be resistant to earthquake shaking. The instructor will provide a simple earthquake simulation table and will shake your structure to see how it withstands an earthquake.

ANTICIPATION GUIDE FOR:
HOW TO HELP: DON'T FAN THE FLAMES

Purpose: To identify what you already know about wildfires and fire prevention, to direct and personalize your reading, and to provide a record of what new information you learned.

———————◆———————

Before you read about wildfires and fire prevention in "How to Help: Don't Fan The Flames," examine each statement below and indicate whether you agree or disagree. Be prepared to discuss your reactions to the statements in groups.

- Wildfires occur only in heavily forested areas.
- Wildfires occur only in semi-arid regions.
- Wildfires are easily preventable.
- Landowners are often responsible when fires burn their homes.
- Homeowners must use huge amounts of water to prevent wildfires from damaging their homes.
- The terrain surrounding your home can be a factor in the house surviving during a wildfire.
- Little can be done to an existing home to improve its ability to survive a wildfire.

HOW TO HELP: DON'T FAN THE FLAMES

1 Improve the Roof
Fire-resistant ceramic tiles, slate or composition shingles, and metal sheets provide better protection than wood.

2 Seal Off Openings
Put metal screens over vents and other openings to block embers.

3 Prune Branches
Flames can jump from branches hanging over the roof of the house.

4 Mow the Lawn
Keep grass short and well watered to hinder the spread of flames.

5 Pick Up Debris
Remove leaf litter and pine needles from gutters, dead limbs from around the house.

Mika Grondahl

DON'T FAN THE FLAMES Each year more Americans move into what is called the wildland-urban interface: zones where human habitation meets forests or grasslands. Many wildfires start in such places; under the right conditions, flames or embers can easily spread to nearby homes. Fortunately, there are ways to make a house in those zones less vulnerable to fire, from the choice of building materials to the landscaping decisions. More information is available at firewise.org.

6 Lighten Landscaping
Spacing out trees
and shrubs makes it
harder for flames to
travel and easier for
firefighters to work.

Writing Assignment

After reading the article "How to Help: Don't Fan the Flames" and conducting your own research, write a persuasive paper encouraging your community to become a "firewise community."

In addition to your own ideas and thoughts, please discuss:

- The people you will address this letter to.

- The history of wildfires in your area (for instance Texas had 27,976 wildfires in the 2010–2011 fire season).

- The hazards (fuels) that are common in your area.

- The steps you will take to make your community firewise.

- The steps you will take to make your home and your family safe.

Listed below are a few links to help you in getting started.

Links:

http://www.firewise.org/
http://www.smokeybear.com/
http://www.fs.fed.us/fire/

Career Investigation

You are enjoying a warm summer day and notice smoke on the horizon. You turn on the news and hear that a wildfire is headed towards your neighborhood. You begin to panic; what should you do first?

While it may be too late to do much, except to escape the ensuing firestorm, there is a lot that can be done to make your home or business more fire resistant.

Listed here are several possible careers to investigate. You may also be able to find a career path not listed. Choose three possible career paths and investigate what you would need to do to be ready to fill one of those positions.

Among the items to look for:

- Education—college? What would you major in? Should you have a minor?

- Working conditions—What is the day to day job like? Will you have to be out in the field all day, or is it a desk job? Is the work strenuous? Are you working on a drilling rig at sea, away from your family for 3–6 months at a time? Is the job "hazardous"?

- Pay scale—What is the average "starting pay" for the position you are seeking? Don't be fooled by looking at "average pay," which may include those who have been working for 20+ years.

- Is the job located in the states, or is there a possibility to travel to other parts of the world?

- Are there any other special requirements?

Architect
Architectural Technician or Technologist
Botanist
Bricklayer
Building Surveyor
Building Technician
Carpenter or Joiner
Civil Engineer
Civil Engineering Technician
Construction Manager
Fence Installer
Firefighter
Fire Inspector/Investigator
Forest Fire Prevention Specialist
Fire Captain
Forest Ranger
Land Surveyor
Landscape Manager
Landscape Scientist
Landscaper
Roofer

Rural Surveyor
Scaffolder
Steel Erector
Steel Fixer
Steeplejack or Lightning Conductor Engineer
Stonemason
Structural Engineer
Thermal Insulation Engineer
Tiler
Town Planner
Town Planning Support Staff
Window Fitter
Wood Machinist

Team Building Activity

You will be assigned to a team of four to five students, who will take on the role of a panelist (consisting of homeowners, business owners, civic leaders, fire marshals, etc.) who will conduct a public hearing on the recent threats of wildfires in the area and propose the community become "firewise." While you are presenting, your classmates will serve as townspeople, business owners, and members of the media.

ANTICIPATION GUIDE FOR:
TUNNEL VISION

Purpose: To identify what you already know about transportation and tunnels, to direct and personalize your reading, and to provide a record of what new information you learned.

———————◆———————

Before you read about tunneling through the Swiss Alps in "Tunnel Vision," examine each statement below and indicate whether you agree or disagree. Be prepared to discuss your reactions to the statements in groups.

- The Gotthard Base Tunnel in Switzerland is the longest tunnel in the world.
- Switzerland's Gotthard Base Tunnel is designed for automobiles, trucks, and trains.
- Traveling through the Gotthard Base Tunnel will save approximately 40 kilometers (25 miles) and about 2 hours over the traditional method of traveling over the Gotthard Pass.
- England's Chunnel and Japan's Seiken Tunnel were more difficult to engineer and construct because they were constructed under water.
- There is already an existing tunnel on Gotthard Pass, so the Gotthard Base Tunnel is redundant and a waste of money.
- The towns to be connected by the Gotthard Base Tunnel are relatively small, so there is no need for the tunnel.
- Through the construction phase of the Gotthard Base Tunnel, workers will excavate 25 million tons of rock.
- The Gotthard Base Tunnel will run up to 2 kilometers (1.2 miles) below the surface of the Alps.
- As with all tunnels, the main bore(s) are only a portion of the work to be done.
- The boring of the tunnels progresses at a nearly steady

TUNNEL VISION

By Roff Smith

Slowed traffic exits the Gotthard Tunnel.
© Melissa Farlow/National Geographic Stock

SHOOTING BENEATH THE MOUNTAINS LIKE A RIFLE BARREL,

SWITZERLAND'S GOTTHARD BASE TUNNEL WILL ALLOW TRAINS TO HURTLE BETWEEN ZURICH AND MILAN ON A FLAT COURSE AT 250 KILOMETERS AN HOUR.

He wasn't the first alpine traveler to wish there were another way. For thousands of years, the Alps have been the great barrier to travel and trade on the Continent. To cross them meant a drawn-out, often perilous trip, or at the very least a tough uphill slog.

Not for much longer. For the past nine years, an army of tunnelers has toiled deep in the hard granitic core of the mighty mountain range known as the Gotthard Massif, constructing the world's longest and deepest railway tunnel.

Last October miners burrowing north from the Italian-speaking canton of Ticino met their counterparts who'd been boring south from German-speaking Sedrun to conclude the digging phase of one bore of the double-barreled tunnel—a handshake moment televised live on Swiss TV and broadcast across Europe. In April, the final breakthrough is expected in the second tube.

At 57 kilometers long, the Gotthard Base Tunnel will handily outstrip the 50-kiilometer Channel Tunnel between England and France and the present record-holder, Japan's

No one has ever tunneled so deep into a mountain, or to such transforming effect.

54-kilometer-long Seikan Tunnel. It will stand alone on the engineering front as well. Where its nearest rivals both pass beneath relatively shallow bodies of water, the Gotthard cuts through the complex basement rocks of a giant, heavily folded mountain range. No one has ever tunneled so deep into a mountain, or to such transforming effect.

When it opens, in 2017, it will render Switzerland, for railroading purposes, as flat as Holland. High-speed passenger trains southbound from Zurich will race along a nearly level course all the way to Milan, booming through the Swiss countryside as fast as 250 kmh, flashing in one side of the mountains and popping out the other a few minutes later. It will be as though the Alps didn't exist. Travel time between the cities will drop from nearly four hours to just over two and a half—quicker and more direct than if you flew.

Beating out the airlines isn't why the Swiss are spending $10 billion on the tunnel: They're

Adapted from "Tunnel Vision" by Roff Smith: National Geographic Magazine, March 2011.

doing it to shift freight, and to curb the spiraling number of trucks clogging their highways and rumbling through their fragile alpine backyard.

Truck traffic has grown exponentially in the sleek new borderless Europe, especially in the Alps, which straddles the fast-growing economic regions of southern Germany and Italy's industrial north. Quiet, neutral, traditionally aloof Switzerland has become a main trucking crossroads. More than a million trucks a year travel over its passes on winding mountain highways and through alpine road tunnels primarily designed for holiday traffic in the 1960s.

The solution, the Swiss decided, was to boost the capacity of the railways to handle freight. And the best way to do that would be to write the mountains and their famous passes out of the equation: Run a set of tracks straight through the bottom of the mountains and out the other side. With no gradients to climb, trains could haul loads twice as heavy and travel twice as fast as those using the old alpine railways. The Gotthard tunnel alone will be able to handle 40 million tonnes of cargo a year.

The tunnel begins its journey under the 2,108-meter pass in the quiet hamlet of Erstfeld, plunging into a hillside through twin concrete portals. It doesn't come out again until Bodio, more than 57 kilometers away, having crossed one of the Continent's great divides. They speak German when you go into the tunnel; Italian when you come out. A rainy day in the quiet hamlet of Erstfeld is likely to be a sunny one in Bodio, and vice versa.

The tunnel avoids the highest (and weightiest) peaks. Its sinuous path seeks out the most favorable geology and skirts potential groundwater complications with lakes that dot the surface some 2 kilometers overhead. Five years and $115 million Swiss francs were spent on fieldwork, drilling, soil samples, and a remote sensing survey to map the massif's nooks and crannies to an accuracy of about ten meters.

They speak German when you go into the tunnel; Italian when you come out.

Nothing about the Gotthard project is small. In the course of building the tunnel, workers will excavate 25 million tonnes of rock, enough to fill a freight train stretching from Zurich to New York or, if you're so inclined, to build five life-size replicas of the Great Pyramid. Some of the dross will be dumped in Lake Lucerne to create an offshore nesting area for birds. The better quality stuff will be ground up for concrete to line the passage. In all, some 152 kilometers of tunnel will be dug and lined—two main bores, at 57 kilometers each, plus kilometers of access shafts, emergency escape passages, ventilation ducts, and crossover points, so trains can shift channels when tracks need repair or maintenance.

In keeping with the scale of the enterprise, the machines that do much of the work are huge. The gigantic creeping machine known as "The Worm," which applies the concrete lining to and lays the drainage pipes, is nearly 600 meters long. A mere 400 meters long but vastly more powerful are four 10-meter diameter Tunnel Boring Machines—TBMs in tunneling parlance. In a typical day each of these 2,700-tonne behemoths will gouge from 20 to 25 meters of solid rock, securing newly dug lengths of tunnel with bolts, shotcrete and steel mesh. Each day they'll consume enough electricity to power 5,000 average suburban homes. Like ships, they have names: Sissi, Heidi, Gabi 1 and Gabi 2.

These girlish monikers, together with a few shrines to St. Barbara, patron saint of miners, represent the distaff side of humanity down here. Of the 2,000 or so miners working on the Gotthard Base Tunnel, none are women.

Over the years the sectors linked up, one by one, and with astonishing precision. When the Gabi 1, burrowing south from

Erstfeld, reached Amsteg, in June 2009, it was a mere five millimeters off course.

The current tunnel builders are following in a long history of Swiss engineering achievement at the Gotthard Massif. Back in the 13th century, a medieval stone mason succeeded in throwing an arched span over the fearsome Schöllenen Gorge that guards the approach to the pass, and with it opened a lucrative trade route into Lombardy.

Later generations came to see the sights. By the early 1800s the road had to be widened and a sturdier bridge built to accommodate the horse-and-carriage trade of the Grand Tour. The poet William Wordsworth, who passed through on his way to Italy in 1820, bemoaned the "arbitrary, pitiless, godless wretches who have removed nature's landmarks by cutting roads through Alps…"

Little could he have imagined what was to follow. By 1870 the railway had arrived, bringing modernity in its wake. It was a monumental undertaking that included blasting a tunnel 15 kilometers long through the tough granite shoulder of one of the Continent's mightiest ranges.

That 19th-century tunnel took ten years to dig and cost at least 199 lives. Louis Favre, the brilliant Swiss engineer who built it, died of a stroke while inspecting the work, aged 53, only months before the tunnel was complete. The inaugural train ran through in 1882, passengers sipping champagne as they covered the distance between Milan and Lucerne in ten hours. Within the first year, a quarter of a million people had taken the trip. And still the numbers grew. A century later, in 1980, another tunnel was opened through the pass, this time for automobiles—at 16.9 kilometers the world's longest road tunnel at the time. A narrow, two-lane affair, it was never designed for trucks, but they came all the same. Gotthard Pass has always been one of the most direct north-south routes through the Alps.

Only one final stretch of virgin rock remained to be dug on the cold and snowy March afternoon when I accompanied AlpTransit's chief engineer, Heinz Ehrbar, on a site visit. By that time, a mere 2.4 kilometers of granite and gneiss separated the miners tunneling south from the ski village of Sedrun from those boring north from Faido.

For Ehrbar, Sedrun was where it all began, fifteen years ago, when he was offered the job of project managing this portion of the tunnel. Although he has since been promoted to chief engineer for the whole project, the sector starting from Sedrun remains special— "Heinz's baby," as one of AlpTransit's surveyors put it—not just for old time's sake, but because this was the toughest stretch, where the geology was the most fickle and complex and the tunnel ran the deepest, up to 2,450 meters below the mountaintops.

"I enjoyed it," he confided as we donned safety gear—boots, hardhat, miner's lamp, high-visibility overalls and a knapsack with a half-hour's supply of oxygen. "A TBM is an impressive piece of machinery but sitting in an operator's chair, watching dials, isn't as satisfying as blasting your way through the rock."

And some of it really needed working. There was no hope of driving a TBM through the rock in the Sedrun sector. Every meter had to be won the old-fashioned way, by blasting or excavating with conventional machines and shoring up. One 1,100-meter-long section of a deformed gneiss known as Kakirit took three years to go through—a rate that would have seen the tunnel as a whole take a century to dig. It was nightmarish stuff for tunneling, buttery soft, prone to collapse, and lacking any structural integrity. To keep the immense weight of the mountain from warping the tunnel out of shape, Ehrbar enlarged the passage, then shored it up with huge constricting steel rings that would give slightly under pressure, slowly easing the walls and ceiling into the desired shape and size.

Just getting to the diggings from Sedrun is an adventure. To reach the bowels of the massif, miners had to burrow into a nearby mountainside, then drill a pair of shafts down 800 meters—twice as deep as the Empire State Building is tall—and install lifts, one to carry workers and materials up and down; the other for heavy equipment. Engineers from South Africa's famously deep gold mines were flown in to sink the shafts.

It's an exhilarating ride, a whooshing plunge in a steel cage with dust and wind whipping all around. "Better than taking the stairs though," Ehrbar quipped as we stepped out at the bottom into a hot damp subterranean world. We hung our jackets on iron pegs hammered into the rock and boarded a jolting, squealing narrow-gauge miner's train for the long ride to the excavation front. They were between blasts when we arrived, the bucket loaders scooping up debris in one bore of the tunnel, the blasting team prepping the face of the other, pumping explosives into a hundred drill holes, and wiring up the charges.

In a few hours they'd blast again, and push the passage another three meters deeper. Meanwhile somewhere on the other side of the rock face, about 2.4 kilometers away, Sissi and Heidi were clawing ever closer through the rock.

On the way back to the lift, Ehrbar pointed out the section they'd labored over, meter by meter, for those three years. "When this is all finished," he said, "I want to go on a test ride through it. They tested the Lötschberg at 288 kilometers an hour. We've got a longer tunnel. I want to speed through here like that."

A swift smooth ride through the Alps in comforting darkness: Adam of Usk would have loved it.

---

you would need to do to be ready to fill one of those positions.

Among the items to look for:

- Education—college? What would you major in? Should you have a minor?

- Working conditions—What is the day to day job like? Will you have to be out in the field all day, or is it a desk job? Is the work strenuous? Are you working on a drilling rig at sea, away from your family for 3–6 months at a time? Is the job "hazardous"?

- Pay scale—What is the average "starting pay" for the position you are seeking? Don't be fooled by looking at "average pay," which may include those who have been working for 20+ years.

- Is the job located in the states, or is there a possibility to travel to other parts of the world?

- Are there any other special requirements?

Civil Engineer
Civil Engineering Technician
Construction Manager
Construction Operative
Construction Plant Mechanic
Construction Plant Operator
Electrician
Electricity Distribution Worker
Engineering Construction Craft Worker
Engineering Construction Technician

Geoscientist
Geo-technician
Glazier
Heating and Ventilation Engineer
Land Surveyor
Measurement and Control Technician
Mechanical Engineering Technician
Plumber
Project Manager
Public Relations Officer
Quarry Operative
Refrigeration/Air Conditioning Engineer
Research Scientist
Risk Management Supervisor
Road Worker
Safety and Health Supervisor
Shift Engineer
Site Supervisor
Structural Engineer
Traffic Engineer
Tunnel Manager

Team Building Activity

You will be assigned to a team of four students, and your team will create a board game modeled after "The Game of Life" to represent choices that would have to be made when deciding to travel through the Gotthard Tunnel or travel overland. Choices within the board game will depend on whether the player chooses to travel by tunnel under the Alps or travel the traditional route over the Alps, and there should be appropriate rewards and consequences for both routes.

ANTICIPATION GUIDE FOR:
THE NEW GREAT WALLS

Purpose: To identify what you already know about architecture, design, and China; to direct and personalize your reading; and to provide a record of what new information you learned.

———————◆———————

Before you read about design and building techniques in "The New Great Walls," examine each statement below and indicate whether you agree or disagree. Be prepared to discuss your reactions to the statements in groups.

- China is the most populous country in the world.
- China has the world's second largest economy.
- China has built structures that can be seen from space.
- China has a higher average income than the United States.
- China spent forty billion dollars on construction for the 2008 Olympics.
- China has maintained its traditional architecture and design during the past ten years of building.
- China's buildings of the past ten years have been designed by China's own architects.
- China's general population is happy with the building boom of the past ten years.
- The Chinese government should not be prevented from taking over someone's property to build new buildings, much like it is not allowed in the United States.
- While many of the foreign architects build from the ground up, several Chinese architects will utilize a vacant building and repurpose it to meet the needs of their client.
- One of China's premiere buildings is a set of towers that lean at 6° towards each other with a suspended section connecting the two towers above the 30th floor, creating a continuous loop effect on the building.

THE NEW GREAT WALLS

By Ted C. Fishman

National Stadium
Architect: Herzog & de Meuron, Switzerland
Completed 2008
Created by Fernando G. Baptista, NG Staff

A view of the Bird's Nest, the Olympic stadium in central Beijing.

© Sean Gallagher/National Geographic Stock

Workers dismantle a protective fence aroun
the Olympic Stadium area.

WITH THE OLYMPICS LOOMING,
CHINA IS PUSHING ARCHITECTURE TO ITS LIMITS
FOR A GIANT COMING-OUT PARTY.

We didn't design it to be Chinese. It's an object for the world.

Lunch for the workers constructing the China World Trade Center Tower in Beijing begins at 11:45. Thousands of hard-hatted men pour out of the site of the 74-story high-rise that will be the city's tallest. Most dig into their lunches on the sidewalk. Others head for a food stand where a tin bowl of sheep-gut soup costs 14 cents. Mr. Wang, who comes from a rural village in Henan Province, runs a crew installing ventilation shafts in the first 30 floors of the trade tower. His helmet, too narrow for his formidable head, sits high and rocks when he talks, more so when he laughs. Wang, at 51, has a burly body and a confident eye, but several of his charges are teenagers fresh off the farm. As boss, he bears responsibility for their mistakes, so sometimes he speeds their training with his boot.

Wang and his crew are part of an army of largely unskilled workers, more than a million strong, that has helped turn Beijing into what is perhaps the largest construction zone in history, with thousands of new projects under way. Once a flat cityscape dominated by the imperial Forbidden City and monumental but drab public buildings, Beijing has been struck by skyscraper fever. Over the past 30 years, China's economy has averaged nearly 10 percent annual GDP growth, driven by the marriage of world-class technology with a vast low-cost workforce. That same dynamic has turned China into an architects' playground, first in Shanghai in the 1990s as its skyline filled in with high-rise marvels, and now in Beijing, which is building at a mad pace in preparation for the 2008 Summer Olympic Games in August.

Beijing's newest buildings push aesthetic and technological bounds, each outshimmering the last. Most major projects have been designed by foreign architects: Chinese clients crave innovation and hunt beyond China to get it, says American architect Brad Perkins, founder of Perkins Eastman in New York. During Mao's Cultural Revolution, architects were more technicians than artists (even the term architect was considered bourgeois), and private architectural firms *(Continued on page 31)*

Adapted from "The New Great Walls" by Ted C. Fishman: National Geographic Magazine, May 2008.

(Continued from page 29) were a rarity until a decade ago. "By turning to foreigners like me," says Perkins, "the Chinese are buying 30 to 40 years of experience they didn't have."

China's low-wage workers in turn allow foreign architects to design structures that would be too costly to build at home, with decorative tops, intricate latticework, and bold engineering. The linear grace of the China World Trade Center Tower, for instance, comes from an innovative cross-bracing system that gives it strength against the city's seismic rumblings and high winds, and from glass louvers engineered to make the most of the city's sunlight. But the tower's architects, Skidmore, Owings & Merrill, also used technology that could be handled by crews working at breakneck speed. The building's prefabricated window walls can be snapped together rather than cut on-site, as they would be with more highly trained workers. Using huge construction crews that work around the clock, foreign architects get to see big projects to completion in China in a remarkably short time, often within three to four years. "Some people in China—including Chinese architects—believe their country has become the Western architects' weapons testing ground," says Perkins.

For centuries China's leaders have reshaped the capital to showcase their power and reflect their preoccupations. The Forbidden City was constructed during the 15th century to project the Ming dynastic rulers' connection to heaven. A throng of Soviet-style halls, stadiums, and vast boulevards sprang up in the 1950s and '60s following the Communist Party's rise to signify the collective strength of workers and the absolute control of Mao's rule. Today Beijing, the national emblem, is being remade as China's global city. When new buildings open, officials like to speak of how the structures embody the country's "soft power." Outsiders, goes the message, need not fear China as an aggressor nation or military power.

This message is clearest in the 40-billion-dollar building spree occasioned by the Olympics, the nation's coming-out party. The buildings say that China is big and powerful, but also inventive, sophisticated, and open. Look at three of the most prominent new structures: One is a stadium that looks like a bird's nest, another an aquatic center that resembles a blue bubbly cube, a third an arts center in the form of an egg as big as a city block. Nests, eggs, and bubbles— a whimsical, approachable China. And then there's the "twisted doughnut," the stunning giant home to CCTV, China's government-run broadcaster. Still unfinished, the building connects at the top with cantilevered sections that meet 531 feet high in the air. Practical-minded Beijingers crane their necks and wonder aloud whether the skewed tower will tumble.

A complaint often heard: Many of these structures are designed for foreign tastes, not Chinese. "China is not confident of its own designs, and people prefer to try something new," observes Du Xiaodong, editor of *Chinese Heritage* magazine in Beijing. "The results are disconnected from whatever's next door, and the newest building in the world sits next to some of the oldest, standing together like strangers."

One of the public shames of Beijing is that its building boom has destroyed most of the city's old *hutong* neighborhoods of traditional courtyard houses, whose residents are often forcibly relocated to make way for projects that enrich local officials and developers. Pei Zhu and Tong Wu, the Chinese architects who designed the digital command center for the Olympics, are among the few architects trying to preserve and adapt what remains of the old city. Instead of razing and building over historic neighborhoods, they'll take a factory constructed during Mao's time and refashion it with courtyards and glass walls that offer vistas of the old city. The approach restores Beijing as a city for walkers. Above all it balances the old with the new, a fitting combination for an ancient capital in transition.

As for Mr. Wang, he will likely be among the million or more migrants who will have returned home or moved on to other jobs before the Olympics commence. When the television cameras roll, the city's futuristic vista will have little place for the workers who built it.

Writing Assignment

After reading the article "The New Great Walls" and conducting your own research, write an expository paper that examines the architecture and design boom over the past ten years in China.

In addition to your own ideas and thoughts, please discuss:

- What has happened to the pace of construction since 2008?

- The occupancy/usage rate of the buildings constructed for the 2008 Olympics.

- How do "the people" feel about the new buildings?

- Should the "old" buildings and architecture be discarded for the sake of the new buildings and architecture?

Listed below are a few links to help you in getting started.

Links:

China's Architectural Curiosities
http://www.inman.com/news/2011/11/3/chinas-architectural-curiosities

Is China's Architectural Ambition Leaving Its Own Talent Behind?
http://www.time.com/time/world/article/0,8599,2064794,00.html

Ancient Chinese Culture
http://www1.chinaculture.org/created/created_ancient.html

Three Stages of China's Architectural History
http://arts.cultural-china.com/en/83Arts4779.html

Building the American Dream in China
http://www.nytimes.com/2012/03/18/magazine/architects-in-china-building-the-american-dream.html?pagewanted=all

Career Investigation

Listed here are several possible careers to investigate. You may also be able to find a career path not listed. Choose three possible career paths and investigate what you would need to do to be ready to fill one of those positions.

Among the items to look for:

- Education—college? What would you major in? Should you have a minor?

- Working conditions—What is the day to day job like? Will you have to be out in the field all day, or is it a desk job? Is the work strenuous? Are you working on a drilling rig at sea, away from your family for 3–6 months at a time? Is the job "hazardous"?

- Pay scale—What is the average "starting pay" for the position you are seeking? Don't be fooled by looking at "average pay," which may include those who have been working for 20+ years.

- Is the job located in the states, or is there a possibility to travel to other parts of the world?

- Are there any other special requirements?

Architect
Architectural Technician or Technologist
Bricklayer
Building Surveyor

Building Technician
Carpenter or Joiner
Carpet Fitter/Floor Layer
Ceiling Fixer
Civil Engineer
Civil Engineering Technician
Construction Contracts Manager
Construction Manager
Construction Operative
Electrician
Electricity Distribution Worker
Engineering Construction Technician
Geoscientist
Geotechnician
Heating and Ventilation Engineer
Insulation Installer
Psychologist
Quantity Surveyor
Landscape Scientist
Landscaper
Mechanical Engineer
Mechanical Engineering Technician
Paint Sprayer
Painter and Decorator
Planning and Development Surveyor
Plasterer
Plumber
Project Manager
Refrigeration/Air Conditioning Technician
Engineer
Roofer

Sheet Metal Worker
Steel Erector
Steel Fixer
Steeplejack or Lightning Conductor
Engineer
Stonemason
Structural Engineer
Tiler
Town Planner
Town Planning Support Staff
Water Network Operative
Welder
Window Fitter

Team Building Activity

You will be assigned to a team of four students. You and your team will assume the roles of architects and city officials to present a news conference (minimum five minutes, maximum ten minutes) to announce the construction of several buildings to house Olympic venues. The Olympic site also happens to be a historic site and will require the demolition or relocation of many historic buildings as well as thousands of businesses and residents. The remaining class members will act as reporters, residents, and other concerned citizens and should be prepared to interact accordingly.

ANTICIPATION GUIDE FOR: SAVING ENERGY: IT STARTS AT HOME

Purpose: To identify what you already know about energy conservation, to direct and personalize your reading, and to provide a record of what new information you learned.

◆

Before you read about energy conservation in "Saving Energy: It Starts at Home," examine each statement below and indicate whether you agree or disagree. Be prepared to discuss your reactions to the statements in groups.

- The average United States household produces 150 pounds of carbon dioxide daily.
- The average European household produces less than 75 pounds of carbon dioxide daily.
- The average United States household produces five times more carbon dioxide than the global average.
- Carbon dioxide is the primary gas connected with the process of "global warming."
- Carbon dioxide emissions have risen 30 percent since 1990.
- Carbon dioxide is a harmful gas and should be eliminated.
- A single gallon of gasoline adds 19.6 pounds of carbon dioxide to the atmosphere.
- Vampire electronics can account for eight percent of a household electric bill.
- Industrial businesses such as refineries, paper plants, etc. are responsible for emitting 20 percent of the United States carbon dioxide.
- To stabilize global warming effects, it is recommended that we reduce our carbon dioxide emissions by 80 percent. This is an impossible number to reach.
- It is recommended to take a "cold turkey" approach and reduce your household emissions immediately.

SAVING ENERGY: IT STARTS AT HOME

By Peter Miller

Photographs by Tyrone Turner

Thermographic photography offers clues to where energy is being wasted in this older house in Connecticut. Red and yellow patches indicate escaping heat, while new double-pane windows appear cool blue. By sealing in warmth, the windows cut heating costs, which can account for up to half a family's energy bill.

*"We're farm people," says Janice Haney of Greensburg, Kansas.
"I enjoy hanging clothes out. We don't have to waste electricity
on the dryer. The good old Kansas wind can do it on its own."*

WE ALREADY KNOW THE FASTEST, LEAST EXPENSIVE
WAY TO SLOW CLIMATE CHANGE: USE LESS ENERGY.

WITH A LITTLE EFFORT, AND NOT MUCH MONEY,

MOST OF US COULD REDUCE OUR ENERGY DIETS BY
25 PERCENT OR MORE—DOING THE EARTH A FAVOR
WHILE ALSO HELPING OUR POCKETBOOKS. SO WHAT'S
HOLDING US BACK?

How close could we come to a lifestyle the planet could handle?

Not long ago, my wife, PJ, and I tried a new diet—not to lose a little weight but to answer a nagging question about climate change. Scientists have reported recently that the world is heating up even faster than predicted only a few years ago, and that the consequences could be severe if we don't keep reducing emissions of carbon dioxide and other greenhouse gases that are trapping heat in our atmosphere. But what can we do about it as individuals? And as emissions from China, India, and other developing nations skyrocket, will our efforts really make any difference?

We decided to try an experiment. For one month we tracked our personal emissions of carbon dioxide (CO_2) as if we were counting calories. We wanted to see how much we could cut back, so we put ourselves on a strict diet. The average U.S. household produces about 150 pounds of CO_2 a day by doing common-place things like turning on air-conditioning or driving cars. That's more than twice the European average and almost five times the global average, mostly because Americans drive more and have bigger houses. But how much should we try to reduce?

For an answer, I checked with Tim Flannery, author of *The Weather Makers: How Man Is Changing the Climate and What It Means for Life on Earth*. In his book, he'd challenged readers to make deep cuts in personal emissions to keep the world from reaching critical tipping points, such as the melting of the ice sheets in Greenland or West Antarctica. "To stay below that threshold, we need to reduce CO_2 emissions by 80 percent," he said.

"That sounds like a lot," PJ said. "Can we really do that?"

It seemed unlikely to me too. Still, the point was to answer a simple question: How close could we come to a lifestyle the planet could handle? If it turned out we couldn't do it, perhaps we could at least identify places where the diet pinched and figure out ways to adjust. So we agreed to shoot for 80 percent less than the U.S. average, which equated to a daily diet of only 30 pounds of CO_2. Then we set out to find a few neighbors to join us. (Continued on page 41)

Adapted from "Saving Energy: It Starts at Home" by Peter Miller: National Geographic Magazine, March 2009.

THE MISSING POWER PLANT
*Instead of building a new 730-megawatt facility like the
Decker Power Plant, the Austin, Texas, electric utility reduced
demand by the same amount through rebates on energy-saving
appliances and other programs. "Go into any store in Austin,
and you can't buy an inefficient air conditioner," says general
manager Roger Duncan. "They just stopped stocking them."*

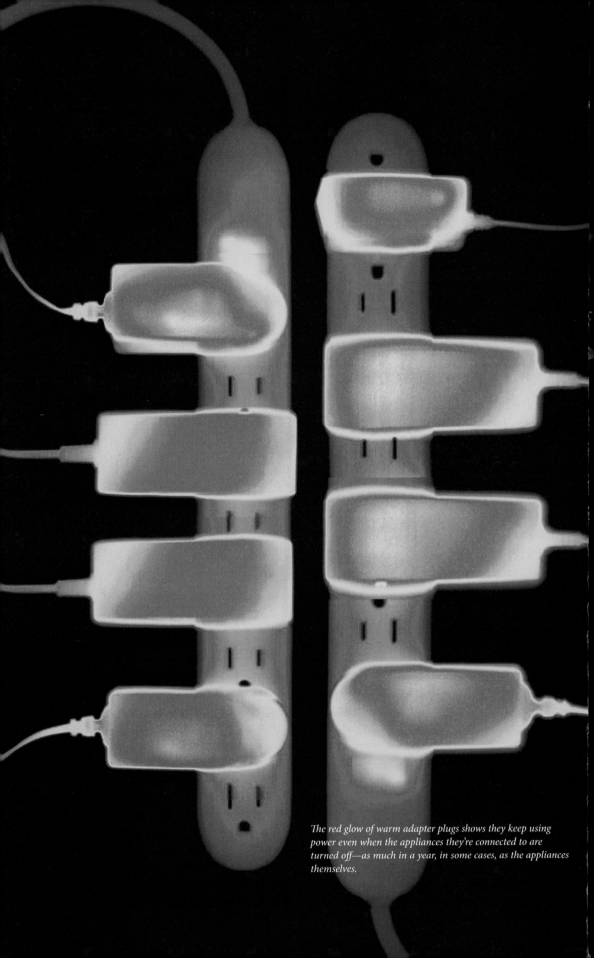

The red glow of warm adapter plugs shows they keep using power even when the appliances they're connected to are turned off—as much in a year, in some cases, as the appliances themselves.

(Continued from page 37) John and Kyoko Bauer were logical candidates. Dedicated greenies, they were already committed to a low-impact lifestyle. One car, one TV, no meat except fish. As parents of three-year-old twins, they were also worried about the future. "Absolutely, sign us up," John said.

Susan and Mitch Freedman, meanwhile, had two teenagers. Susan wasn't sure how eager they would be to cut back during their summer vacation, but she was game to give the diet a try. As an architect, Mitch was working on an office building designed to be energy efficient, so he was curious how much they could save at home. So the Freedmans were in too.

We started on a Sunday in July, an unseasonably mild day in Northern Virginia, where we live. A front had blown through the night before, and I'd opened our bedroom windows to let in the breeze. We'd gotten so used to keeping our air-conditioning going around the clock, I'd almost forgotten the windows even opened. The birds woke us at five with a pleasant racket in the trees, the sun came up, and our experiment began.

Our first challenge was to find ways to convert our daily activities into pounds of CO_2. We wanted to track our progress as we went, to change our habits if necessary.

PJ volunteered to read our electric meter each morning and to check the odometer on our Mazda Miata. While she was doing that, I wrote down the mileage from our Honda CR-V and pushed my way through the shrubs to read the natural gas meter. We diligently recorded everything on a chart taped to one of our kitchen cabinets. A gallon of gasoline, we learned, adds a whopping 19.6 pounds of CO_2 to the atmosphere, a big chunk of our daily allowance. A kilowatt-hour (kWh) of electricity in the U.S. produces 1.5 pounds of CO_2. Every 100 cubic feet of natural gas emits 12 pounds of CO_2.

To get a rough idea of our current carbon footprint, I plugged numbers from recent utility bills into several calculators on websites. Each asked for slightly different information, and each came up with a different result. None was flattering. The Environmental Protection Agency (EPA) website figured our annual CO_2 emissions at 54,273 pounds, 30 percent higher than the average American family with two people; the main culprit was the energy we were using to heat and cool our house. Evidently, we had further to go than I thought.

I began our campaign by grabbing a flashlight and heading down to the basement. For most families, the water heater alone consumes 12 percent of their house's energy. My plan was to turn down the heater's thermostat to 120°F, as experts recommend. But taking a close look at our tank, I saw only "hot" and "warm" settings, no degrees. Not knowing what that meant exactly, I twisted the dial to warm and hoped for the best. (The water turned out to be a little cool, and I had to adjust it later.)

When PJ drove off in the CR-V to pick up a friend for church, I hauled out gear to cut the grass: electric lawn mower, electric edger, electric leaf blower. Then it dawned on me: All this power-sucking equipment was going to cost us in CO_2 emissions. So I stuffed everything back into the garage, hopped in the Miata, and buzzed down the street to Home Depot to price out an old-fashioned push reel mower.

The store didn't have one, so I drove a few miles more to Lawn & Leisure, an outfit that specializes in lawn mowers. They were out too, though they had plenty of big riding mowers on display. (The average gasoline-powered push mower, I'd learned, puts out as much pollution per hour as eleven cars—a riding mower as much as 34 cars.) My next stop was Walmart, where I found another empty spot on the rack. I finally tried Sears, which had one manual mower left, the display model.

I'd seen advertisements for the latest reel mowers that made them sound like precision instruments, not the clunky beast I pushed as a teenager. But when I gave the display model a spin across the sales floor, I was disappointed. The reel felt clumsy compared with my corded electric model, which I can easily maneuver with one hand. I got back in the car empty-handed and drove home.

As I pulled into the driveway, I had the sinking realization I'd been off on a fool's errand. I didn't know exactly how foolish until the next morning, when we added up the numbers. I'd driven 24 miles in search of a more Earth-friendly mower. PJ had driven 27 miles to visit a friend in an assisted-living facility. We'd used 32 kWh of electricity and 100 cubic feet of gas to cook dinner and dry our clothes. Our total CO_2 emissions for the day: 105.6 pounds. Three and a half times our target.

"Guess we need to try harder," PJ said.

We got some help in week two from a professional "house doctor," Ed Minch, of Energy Services Group in Wilmington, Delaware. We asked Minch to do an energy audit of our house to see if we'd missed any easy fixes. The first thing he did was walk around the outside of the house, looking at how the "envelope" was put together. Had the architect and builder created any opportunities for air to seep in or out, such as overhanging floors? Next he went inside and used an infrared scanner to look at our interior walls. A hot or cold spot might mean that we had a duct problem or that insulation in a wall wasn't doing its job. Finally his assistants set up a powerful fan in our front door to lower air pressure inside the house and force air through whatever leaks there might be in the shell of the house. Our house, his instruments showed, was 50 percent leakier than it should be.

One reason, Minch discovered, was that our builder had left a narrow, rectangular hole in our foundation beneath the laundry

I didn't realize how foolish I was until the next morning: I'd driven 24 miles in search of a more Earth-friendly lawn mower.

room—for what reason we could only guess. Leaves from our yard had blown through the hole into the crawl space. "There's your big hit," he said. "That's your open window." I hadn't looked inside the crawl space in years, so there could have been a family of monkeys under there for all I knew. Sealing up that hole was now a priority, since heating represents up to half of a house's energy costs, and cooling can account for a tenth.

Air rushing in through the foundation was only part of the problem, however. Much of the rest was air seeping out of a closet on our second floor, where a small furnace unit was located. The closet had never been completely drywalled, so air filtered through insulation in the roof to the great outdoors. Minch recommended we finish the drywalling when the time comes to replace the furnace.

Minch also gave us tips about lighting and appliances. "A typical kitchen these days has ten 75-watt spots on all day," he said. "That's a huge waste of money." Replacing them with compact fluorescents could save a homeowner $200 a year. Refrigerators, washing machines, dishwashers, and other appliances, in fact, may represent half of a household's electric bill. Those with Energy Star labels from the EPA are more efficient and may come with rebates or tax credits when you buy them, Minch said.

There was no shortage of advice out there, I discovered, about ways to cut back on CO_2 emissions. Even before Minch's visit, I'd collected stacks of printouts and brochures from environmental websites and utility companies. In a sense, there's almost too much information.

"You can't fix everything at once," John Bauer said when I asked how he and Kyoko were getting along. "When we became vegetarians, we didn't do it all at once. First the lamb went. Then the pork. Then the beef.

YOU GET TO READ THE PAPER TOO
Commuters on a Metrorail train contribute only half as much CO_2 to the atmosphere as drivers on the Beltway around Washington, D.C. For every mile on the road, an average American car—often carrying just one or two people—pumps a pound of CO_2 into the sky. Emissions from operating an electric train (mainly from coal-fired power plants) are spread among thousands of riders.

Finally the chicken. We've been phasing out seafood for a few years now. It's no different with a carbon diet."

Good advice, I'm sure. But everywhere I looked I saw things gobbling up energy. One night I sat up in bed, squinted into the darkness, and counted ten little lights: cell phone charger, desktop calculator, laptop computer, printer, clock radio, cable TV box, camera battery recharger, carbon monoxide detector, cordless phone base, smoke detector. What were they all doing? A study by the Lawrence Berkeley National Laboratory found that "vampire" power sucked up by electronics in standby mode can add up to 8 percent of a house's electric bill. What else had I missed?

"You can go nuts thinking about everything in your house that uses power," said Jennifer Thorne Amann, author of Consumer Guide to Home Energy Savings, who had agreed to be our group's energy coach. "You have to use common sense and prioritize. Don't agonize too much. Think about what you'll be able to sustain after the experiment is over. If you have trouble reaching your goal in one area, remember there's always something else you can do."

At this point we left home for a long weekend to attend the wedding of my niece, Alyssa, in Oregon. While we were gone, the house sitter caring for our two dogs continued to read our gas and electric meters, and we kept track of the mileage on our rental car as we drove from Portland (Continued on page 46)

Today's internal combustion engines are inefficient at converting fuel to motion. Cars waste up to 85 percent of the energy from the fuel in their tanks, losing a big chunk as heat.

(Continued from page 43) to the Pacific coast. I knew this trip wasn't going to help our carbon diet any. But what was more important, after all, reducing CO_2 emissions or sharing a family celebration?

That's the big question. How significant are personal efforts to cut back? Do our actions add up to anything meaningful, or are we just making ourselves feel better? I still wasn't sure. As soon as we returned home to Virginia, I started digging up more numbers.

The United States, I learned, produces a fifth of the world's CO_2 emissions, about six billion metric tons a year. That staggering amount could reach seven billion by 2030, as our population and economy continue to grow. Most of the CO_2 comes from energy consumed by buildings, vehicles, and industries. How much CO_2 could be avoided, I started to wonder, if we all tightened our belts? What would happen if the whole country went on a carbon diet?

Buildings, not cars, produce the most CO_2 in the United States. Private residences, shopping malls, warehouses, and offices account for 38 percent of the nation's emissions, mainly because of electricity use. It doesn't help that the average new house in the United States is 45 percent bigger than it was 30 years ago.

Companies like Walmart that maintain thousands of their own buildings have discovered they can achieve significant energy savings. A pilot Supercenter in Las Vegas consumes up to 45 percent less power than similar stores, in part by using evaporative cooling units, radiant floors, high-efficiency refrigeration, and natural light in shopping areas. Retrofits and smart design could reduce emissions from buildings in this country by 200 million tons of CO_2 a year, according to researchers at Oak Ridge

Buildings, not cars, produce the most CO_2 in the U.S. The average new house is 45 percent bigger than it was 30 years ago.

National Laboratory. But Americans are unlikely to achieve such gains, they say, without new building codes, appliance standards, and financial incentives. There are simply too many reasons not to.

Commercial building owners, for example, have had little incentive to pay more for improvements like high-efficiency windows, lights, heating, or cooling systems since their tenants, not they, pay the energy bills, said Harvey Sachs of the American Council for an Energy-Efficient Economy. For homeowners, meanwhile, efficiency takes a backseat whenever money is tight. In a 2007 survey of Americans, 60 percent said they didn't have enough savings to pay for energy-related renovations. If given an extra $10,000 to work with, only 24 percent said they would invest in efficiency. What did the rest want? Granite countertops.

After buildings, transportation is the next largest source of CO_2, producing 34 percent of the nation's emissions. Carmakers have been told by Congress to raise fuel economy standards by 40 percent by 2020. But emissions will still grow, because the number of miles driven in this country keeps going up. One big reason: Developers keep pushing neighborhoods farther into the countryside, making it unavoidable for families to spend hours a day in their cars. An EPA study estimated that greenhouse gas emissions from vehicles could increase 80 percent over the next 50 years. Unless we make it easier for Americans to choose buses, subways, and bikes over cars, experts say, there's little chance for big emissions cuts from vehicles.

The industrial sector represents the third major source of CO_2. Refineries, paper plants, and other facilities emit 28 percent of the nation's total. You would think such enterprises would have eliminated

THE POWERED HOUSE

Electricity is the biggest source of power for U.S. homes—and for every kilowatt-hour used, 2.2 are "lost" as that energy is generated and sent over transmission lines. So, even small changes in our habits can scale up to big reductions in carbon emissions.

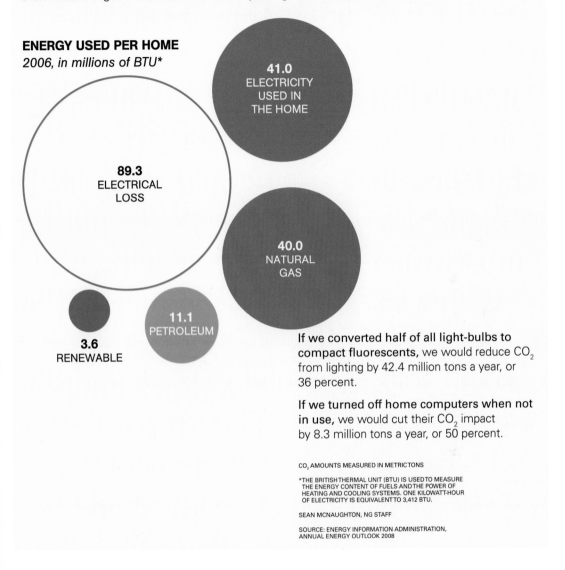

ENERGY USED PER HOME
2006, in millions of BTU*

89.3
ELECTRICAL
LOSS

41.0
ELECTRICITY
USED IN
THE HOME

40.0
NATURAL
GAS

11.1
PETROLEUM

3.6
RENEWABLE

If we converted half of all light-bulbs to compact fluorescents, we would reduce CO_2 from lighting by 42.4 million tons a year, or 36 percent.

If we turned off home computers when not in use, we would cut their CO_2 impact by 8.3 million tons a year, or 50 percent.

CO_2 AMOUNTS MEASURED IN METRIC TONS

*THE BRITISH THERMAL UNIT (BTU) IS USED TO MEASURE THE ENERGY CONTENT OF FUELS AND THE POWER OF HEATING AND COOLING SYSTEMS. ONE KILOWATT-HOUR OF ELECTRICITY IS EQUIVALENT TO 3,412 BTU.

SEAN MCNAUGHTON, NG STAFF

SOURCE: ENERGY INFORMATION ADMINISTRATION, ANNUAL ENERGY OUTLOOK 2008

inefficiencies long ago. But that isn't always the case. For firms competing in global markets, making the best product at the right price comes first. Reducing greenhouse gases is less urgent. Some don't even track CO_2 emissions.

A number of corporations such as Dow, DuPont, and 3M have shown how profitable efficiency can be. Since 1995, Dow has saved seven billion dollars by reducing its energy intensity—the amount of energy consumed per pound of product—and during the past few decades it has cut its CO_2 emissions by 20 percent. To show other companies how to make such gains, the Department of Energy (DOE) has been sending teams of experts into

700 or so factories a year to analyze equipment and techniques. Yet even here change doesn't come easily. Managers are reluctant to invest in efficiency unless the return is high and the payback time is short. Even when tips from the experts involve no cost at all—such as "turn off the ventilation in unoccupied rooms"—fewer than half of such fixes are acted upon. One reason is inertia. "Many changes don't happen until the maintenance foreman, who knows how to keep the old equipment running, dies or retires," said Peggy Podolak, senior industrial energy analyst at DOE.

But change is coming anyway. Most business leaders expect federal regulation of CO_2 emissions in the near future. Already, New York and nine other northeastern states have agreed on a mandatory cap-and-trade system similar to the one started in Europe in 2005. Under the plan, launched last year, emissions from large power plants will be reduced over time, as each plant either cuts emissions or purchases credits from other companies that cut their emissions. A similar scheme has been launched by the governors of California and six other western states and the premiers of four Canadian provinces.

So how do the numbers add up? How much CO_2 could we save if the whole nation went on a low carbon diet? A study by McKinsey & Company, a management consulting firm, estimated that the United States could avoid 1.3 billion tons of CO_2 emissions a year, using only existing technologies that would pay for themselves in savings. Instead of growing by more than a billion tons by 2020, annual emissions in the U.S. would drop by 200 million tons a year. We already know, in other words, how to freeze CO_2 emissions if we want to.

Not that there won't still be obstacles. Every sector of our economy faces challenges, said energy-efficiency guru Amory Lovins of the Rocky Mountain Institute. "But they all have huge potential. I don't know anyone who has failed to make money at energy efficiency.

There's so much low-hanging fruit, it's falling off the trees and mushing up around our ankles."

By the last week in July, PJ and I were finally getting into the flow of the reduced carbon lifestyle. We walked to the neighborhood pool instead of driving, biked to the farmers market on Saturday morning, and lingered on the deck until dark, chatting over the chirping of the crickets. Whenever possible I worked from home, and when I commuted I took the bus and subway. Even when it got hot and humid, as it does in Virginia in July, we were never really uncomfortable, thanks in part to the industrial-size ceiling fan we installed in the bedroom in late June.

"That fan's my new best friend," PJ said.

Our numbers were looking pretty good, in fact, when we crossed the finish line on August 1. Compared with the previous July, we slashed electricity use by 70 percent, natural gas by 40 percent, and reduced our driving to half the national average. In terms of CO_2, we trimmed our emissions to an average of 70.5 pounds a day, which, though twice as much as we'd targeted as our goal, was still half the national average.

These were encouraging results, I thought, until I factored in emissions from our plane trip to Oregon. I hadn't expected that a modern aircraft packed with passengers would emit almost half as much CO_2 per person as PJ and I would have produced if we'd driven to Oregon and back in the CR-V. The round-trip flight added the equivalent of 2,500 pounds of CO_2 to our bottom line, more than doubling our daily average from 70.5 pounds of CO_2 to 150 pounds—five times our goal. So much for air travel.

By comparison, the Bauers did significantly better, though they also faced setbacks. Since their house is all electric, Kyoko Bauer had tried to reduce her use of the clothes dryer by hanging laundry on a rack outside, as she and John had done when they lived in arid

TRANSPORTATION TOLLS

Cars and light trucks consume the lion's share of petroleum used for transportation in the U.S. Modest changes in efficiency and driving habits could add up to significant fuel savings.

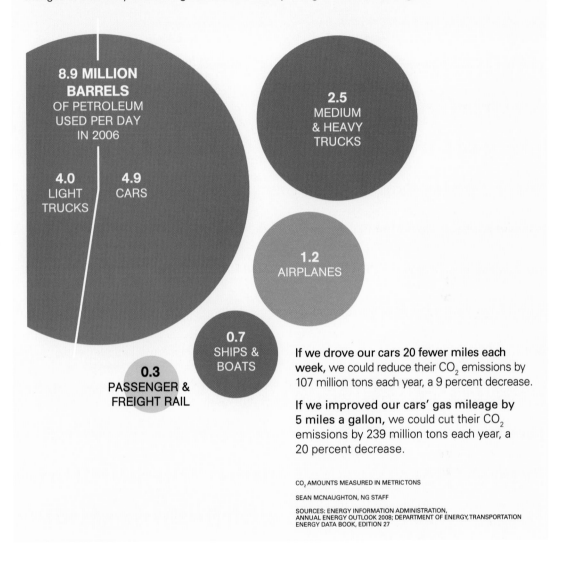

8.9 MILLION BARRELS OF PETROLEUM USED PER DAY IN 2006

4.0 LIGHT TRUCKS

4.9 CARS

2.5 MEDIUM & HEAVY TRUCKS

1.2 AIRPLANES

0.7 SHIPS & BOATS

0.3 PASSENGER & FREIGHT RAIL

If we drove our cars 20 fewer miles each week, we could reduce their CO_2 emissions by 107 million tons each year, a 9 percent decrease.

If we improved our cars' gas mileage by 5 miles a gallon, we could cut their CO_2 emissions by 239 million tons each year, a 20 percent decrease.

CO_2 AMOUNTS MEASURED IN METRIC TONS

SEAN MCNAUGHTON, NG STAFF

SOURCES: ENERGY INFORMATION ADMINISTRATION, ANNUAL ENERGY OUTLOOK 2008; DEPARTMENT OF ENERGY, TRANSPORTATION ENERGY DATA BOOK, EDITION 27

Western Australia. But with their busy three-year-olds, Etienne and Ajanta, she was doing as many as 14 loads a week, and it took all day for clothes to dry in Virginia's humid air. "It wasn't as convenient as I hoped," she said. "I had to race home from shopping a couple of times before it started to rain." Their bottom line: 97.4 pounds of CO_2 a day.

For the Freedmans, driving turned out to be the big bump in the road. With four cars and everyone commuting to a job every day—including Ben and Courtney—they racked up 4,536 miles during the month. "I don't know how we could have driven less," Susan said. "We were all going in different directions and there wasn't any other way to get there." Their bottom line: 248 pounds of CO_2 a day.

When we received our electric bill for July, PJ and I were pleased that our efforts had saved us $190. We decided to use a portion of this windfall to offset (Continued on page 54)

A GREEN DREAM HOUSE
After a monster tornado swept away their home in 2007, Jill and Scott Eller of Greensburg, Kansas, decided to rebuild using a more efficient design. Their new house, constructed from structural insulated panels like the one Jill is holding, is expected to be much more airtight than standard wood-frame houses. As a bonus, the domes should deflect all but the strongest of winds.

BRINGING THE FARM TO THE CITY
If tomatoes, cucumbers, lettuce, strawberries, pumpkins, and other crops can grow on a barge in the Hudson River, then why not on New York City rooftops? That was the idea behind the Science Barge, a prototype of a carbon-neutral hydroponic farm that saves energy by eliminating the need for transportation.

(Continued from page 49) the airline emissions. After doing a little homework, we contributed $50 to Native Energy, one of many companies and nonprofits that save CO_2 by investing in wind farms, solar plants, and other renewable energy projects. Our purchase was enough to counteract a ton of jet emissions, roughly what we added through our trip and then some.

We can do more, of course. We can sign up with our utility company for power from regional wind farms. We can purchase locally grown foods instead of winter raspberries from Chile and bottled water from Fiji. We can join a carbon-reduction club through a neighborhood church, Scout troop, Rotary Club, PTA, or environmental group. If we can't find one, we could start one. "If you can get enough people to do things in enough communities, you can have a huge impact," said David Gershon, author of *Low Carbon Diet: A 30-Day Program to Lose 5,000 Pounds*. "When people are successful, they say, 'Wow, I want to go further. I'm going to push for better public transportation, bike lanes, whatever.' Somebody called this the mice-on-the-ice strategy. You don't have to get any one element to work, but if you come at it from enough different directions, eventually the ice cracks."

W ill it make any difference? That's what we really wanted to know. Our low carbon diet had shown us that, with little or no hardship and no major cash outlays, we could cut day-to-day emissions of CO_2 in half—mainly by wasting less energy at home and on the highway. Similar efforts in office buildings, shopping malls, and factories throughout the nation, combined with incentives and efficiency standards, could halt further increases in U.S. emissions.

That won't be enough by itself, though. The world will still suffer severe disruptions unless humanity reduces emissions sharply—and they've risen 30 percent since 1990. As much as 80 percent of new energy demand in the next decade is projected to come from China, India, and other developing nations. China is building the equivalent of two midsize coal-fired power plants a week, and by 2007 its CO_2 output surpassed that of the U.S. Putting the brakes on global emissions will be more difficult than curbing CO_2 in the United States, because the economies of developing nations are growing faster. But it begins the same way: By focusing on better insulation in houses, more efficient lights in offices, better gas mileage in cars, and smarter processes in industry. The potential exists, as McKinsey reported last year, to cut the growth of global emissions in half.

Yet efficiency, in the end, can only take us so far. To get the deeper reductions we need, as Tim Flannery advised—80 percent by 2050 (or even 100 percent, as he now advocates)—we must replace fossil fuels faster with renewable energy from wind farms, solar plants, geothermal facilities, and biofuels. We must slow deforestation, which is an additional source of greenhouse gases. And we must develop technologies to capture and bury carbon dioxide from existing power plants. Efficiency can buy us time—perhaps as much as two decades—to figure out how to remove carbon from the world's diet.

The rest of the world isn't waiting for the United States to show the way. Sweden has pioneered carbon-neutral houses, Germany affordable solar power, Japan fuel-efficient cars, the Netherlands prosperous cities filled with bicycles. Do Americans have the will to match such efforts?

Maybe so, said R. James Woolsey, former director of the CIA, who sees a powerful, if unlikely, new alliance forming behind energy efficiency. "Some people are in favor of it because it's a way to make money, some because they're worried about terrorism or global warming, some because they think it's their religious duty," he said. "But it's all coming together, and politicians (Continued on page 58)

Blue signifies the cool air escaping as four-year-old Eva Turner dawdles at the fridge. That's not so bad: Today's models use a third less energy than those of 30 years ago.

A CARBON REDUCTION PLAN

By investing in new technology or adopting approaches already available, we could cut U.S. greenhouse gas emissions by three billion tons a year, more than offsetting the increases expected by 2030 as our population and economy grow. And the money saved from efficiencies in how we use energy (below) could help pay for improvements in how we generate energy (right).

KEY SECTORS

1,127 million tons per year

729 million tons per year

520 million tons per year

486 million tons per year

357 million tons per year

CUTS THAT SAVE MONEY

About 40 percent of possible cuts could come from measures that save billions of dollars a year (below). Most of these savings are found in building improvements, such as more efficient lighting, and transportation improvements like better fuel efficiency.

CO₂ REDUCTIONS ▶ (IN BILLIONS OF TONS PER YEAR)

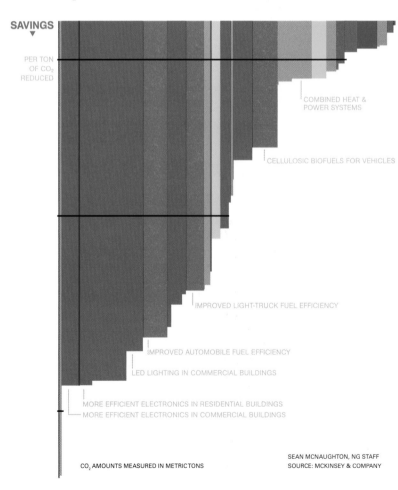

SAVINGS ▼

PER TON OF CO₂ REDUCED

COMBINED HEAT & POWER SYSTEMS

CELLULOSIC BIOFUELS FOR VEHICLES

IMPROVED LIGHT-TRUCK FUEL EFFICIENCY

IMPROVED AUTOMOBILE FUEL EFFICIENCY

LED LIGHTING IN COMMERCIAL BUILDINGS

MORE EFFICIENT ELECTRONICS IN RESIDENTIAL BUILDINGS
MORE EFFICIENT ELECTRONICS IN COMMERCIAL BUILDINGS

CO₂ AMOUNTS MEASURED IN METRIC TONS

SEAN MCNAUGHTON, NG STAFF
SOURCE: MCKINSEY & COMPANY

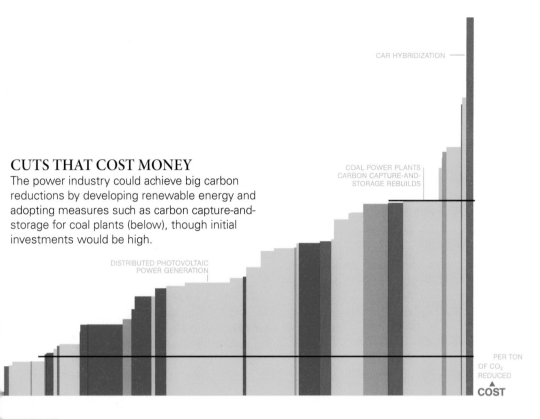

CUTS THAT COST MONEY

The power industry could achieve big carbon reductions by developing renewable energy and adopting measures such as carbon capture-and-storage for coal plants (below), though initial investments would be high.

CAR HYBRIDIZATION

COAL POWER PLANTS
CARBON CAPTURE-AND-
STORAGE REBUILDS

DISTRIBUTED PHOTOVOLTAIC
POWER GENERATION

PER TON
OF CO_2
REDUCED

▲ COST

EDUCTIONS ▶ (IN BILLIONS OF TONS PER YEAR)

A worker in Washington, D.C., installs a triple-glazed window in a structure designed to meet strict "green building" standards. Advanced lighting, heating, cooling, and water systems—as well as a green roof—contribute to a small carbon footprint. This can reduce energy costs by up to 75 percent. But many firms hesitate to invest in efficiency if up-front costs seem too high or payback times too long.

(Continued from page 54) are starting to notice. I call it a growing coalition between the tree huggers, the do-gooders, the sodbusters, the cheap hawks, the evangelicals, the utility shareholders, the mom-and-pop drivers, and Willie Nelson."

This movement starts at home with the changing of a lightbulb, the opening of a window, a walk to the bus, or a bike ride to the post office. PJ and I did it for only a month, but I can see the low carbon diet becoming a habit.

"What do we have to lose?" PJ said.

Writing Assignment

After reading the article "Saving Energy: It Starts at Home" and conducting your own research, write a brief paper discussing the effects of carbon dioxide in our atmosphere and the steps we must take to reduce our carbon dioxide emissions.

In addition to your own ideas and thoughts, please discuss:

- What is meant by "carbon footprint?"

- What is carbon, and why is carbon important in our lives?

- What is carbon dioxide (CO_2), and what are the benefits and detriments of carbon dioxide?

- How has the level of carbon dioxide changed in the past one hundred years?

- What countries are the largest contributors of CO_2 emissions? What countries are the largest contributors of CO_2 emissions per capita? What reason can be given for the significant difference?

Career Investigation

What would you do with an extra $200? What about $500 or even $1000? With some simple lifestyle changes that will reduce greenhouse gases, you can save that much money every year and do a tremendous amount of good for the world.

Listed here are several possible careers to investigate. You may also be able to find a career path not listed. Choose three possible career paths and investigate what you would need to do to be ready to fill one of those positions.

Among the items to look for:

- Education—college? What would you major in? Should you have a minor?

- Working conditions—What is the day to day job like? Will you have to be out in the field all day, or is it a desk job? Is the work strenuous? Are you working on a drilling rig at sea, away from your family for 3–6 months at a time? Is the job "hazardous"?

- Pay scale—What is the average "starting pay" for the position you are seeking? Don't be fooled by looking at "average pay," which may include those who have been working for 20+ years.

- Is the job located in the states, or is there a possibility to travel to other parts of the world?

- Are there any other special requirements?

Architect
Architectural Technician or Technologist
Bricklayer
Building Surveyor
Building Technician
Carpenter or Joiner
Carpet Fitter-Floor Layer
Ceiling Fixer
Chemical Engineer
Chemical Engineering Technician
Chemist
Civil Engineer
Civil Engineering Technician
Construction Manager
Consumer Scientist
Ecologist
Electronics Engineer
Engineering Construction Craftworker
Engineering Construction Technician
Geoscientist
Geotechnician
Heating and Ventilation Engineer
Insulation Installer
Kitchen and Bathroom Fitter
Plasterer
Plumber

Project Manager
Roofer
Sheet Metal Worker
Structural Engineer
Technical Brewer
Thermal Insulation Engineer
Welder
Window Fitter

Team Building Activity

You will be assigned to a team of four students and will design a "green" energy efficient home, with the goal of having a net zero or negative carbon footprint. The home may be single- or multi-story. Your design will be presented to the class as an architect would present designs to his/her clients. All members of the group should participate in the presentation.

Below are a few links to help you get started.

Links:

Home power designs
http://homepower.com/basics/design/

Energy efficient building design
http://www.fsec.ucf.edu/en/publications/html/FSEC-GP-33-88/4-home-design.pdf

Efficient building PDF
http://www.cpo.com/home/portals/2/downloads/pdf/stem/efficient_bldgs.pdf

ANTICIPATION GUIDE FOR:
NEXT: SIMULATING WILDFIRES

Purpose: To identify what you already know about how wildfires occur and spread, to direct and personalize your reading, and to provide a record of what new information you learned.

———————————◆———————————

Before you read "Next: Simulating Wildfires," examine each statement below and indicate whether you agree or disagree. Be prepared to discuss your reactions to the statements in groups.

- Wildfires are less threatening than tornados because you have plenty of warning before a wildfire threatens your home.
- Windborne embers are one of the most threatening aspects of a wildfire.
- Wildfires occur only in heavily forested areas.
- To learn more about wildfires, insurance companies built a test center where they burn houses under wildfire conditions.
- A home in the path of an approaching wildfire can ignite within seconds of the first embers hitting the home.
- Homeowners must use huge amounts of water to prevent wildfires from damaging their homes.
- Many homes that burn during a wildfire actually burn from the inside out.
- Little can be done to an existing home to improve its ability to survive a wildfire.

NEXT: SIMULATING WILDFIRES

By Luna Shyr

Built for burning, a test house (left) faces a simulation of an ember storm typical of wildfires.

1. To generate the fiery bits of material, bins of bark mulch soaked in lighter fluid are set aflame.

2. Fans blow embers up metal tubes and at the house, simulating winds of 10 to 20 miles an hour.

3. The house sits on a turntable and can be fitted with different roofs, sidings, windows, and vegetation.

The simulation showed that embers can easily ignite debris like pine needles caught in a house gutter.

© Joel Sartore/National Geographic Stock

RESEARCHERS
SEEK THE WEAK SPOTS
IN A HOME'S DEFENSES.

Call it playing with fire for a purpose.

Four decades of studying fires have led Jack Cohen of the U.S. Forest Service to one conclusion: When it comes to wildfires, the greatest threat to homes isn't from walls of flame sweeping through residential areas. It's from the houses themselves—their construction, materials, even landscaping—and their susceptibility to embers, the tiny bits of burning material he calls firebrands.

Cohen has seen thousands of homes succumb to fire, including some of the approximately 5,500 consumed in the California infernos of 2003 and 2007. The following year the Department of Homeland Security agreed to fund development of software that will eventually enable homeowners and fire agencies to evaluate vulnerabilities in houses and other structures. This, says Cohen, is a vital step toward preventing disaster. To prove his point, he's enlisted an impressive tool: a full-scale house that can be set afire, refitted with different materials, and then set aflame again.

Call it playing with fire for a purpose. The simulations take place in a giant facility situated on 90 acres in the South Carolina countryside. Here the Insurance Institute for Business & Home Safety, with funding from some 60 insurance companies, recreates the conditions of wildfires, hurricanes, and the like in order to study their impact on buildings and to develop protection guidelines. "There's nothing else like this lab," says President and CEO Julie Rochman. "Our number one obsession is that the science be right."

The challenges are enormous. Fire chiefs and forestry experts attest to the scientific accuracy of the fire simulations, but in the course of that achievement, ember machines have burst into flames, and metal pipes have buckled. The 105 "wind" fans devour so much energy that the nearly year-old facility has its own power substation. The tests, however, have yielded valuable information that is documented on video and in photographs.

To isolate vulnerable spots on a building in the midst of a blaze, the 1,400-square-foot test house is bombarded with embers generated by igniting bins of mulch. The structure can be

Adapted from "Next: Simulating Wildfires" by Luna Shyr: National Geographic Magazine, September 2011.

Protecting Your Home

1 **Roof** Install class A-rated coverings. Shape and age also affect a roof's vulnerability to fire.

2 **Windows** Dual-pan tempered glass is m resistant to breakag from heat exposure

Jason Lee

fitted with different kinds of siding, windows, gutters, and roofs. Among the lessons learned: Vinyl gutters readily melt, and embers can infiltrate homes through vents, windows, and roofs. "We were a little surprised how quickly things happened once embers blew onto the roof," says Rochman. "We saw ignition in seconds."

That's the point Cohen hopes the software based on his research will drive home. "When

ers Keep gutters **4 Plants** Remove dry,
of combustible dead vegetation. Prune
is such as pine overhanging trees or
les and leaves. branches.

wildfires burn intensely, they produce millions of firebrands that come down like a blizzard," he says. Once inside a house, they can potentially burn it from the inside out. The software will help users pinpoint areas prone to igniting.

At the lab, meantime, another test will look at how radiant heat from a burning structure can cause its neighbor to combust. And on deck as the next great simulation challenge: creating the perfect hailstorm.

Writing Assignment

After reading the article "Next: Simulating Wildfires" and conducting your own research, write a brief paper discussing methods researchers use to simulate and study wildfires, how they spread, and methods to prevent them.

In addition to your own ideas and thoughts, please discuss:

- Specific steps homeowners can implement to protect their home/ property from an approaching wildfire.

- Is it possible to design/build a truly fireproof home? If so, is the design realistic?

- Reflect on your current living arrangements; are there steps you should take to further prevent the spread of wildfires?

Listed below are a few links to help you in getting started.

Links:

South Carolina Forestry
http://www.state.sc.us/forest/fire.htm

South Carolina Forestry Commission
http://www.state.sc.us/forest/refwild.htm

Wildfire Training Newscast
http://www.wistv.com/Global/story.asp?S=14317470

Disaster Safety
http://www.disastersafety.org/Wildfire

Protecting your home
http://www.nifc.gov/prevEdu/prevEdu_main.html

Career Investigation

You are enjoying a warm summer day and notice smoke on the horizon. You turn on the news and hear that a wildfire is headed towards your neighborhood. You begin to panic; what should you do first?

While it may be too late to do much, except to escape the ensuing firestorm, there is a lot that can be done to make your home or business more fire resistant.

Listed here are several possible careers to investigate. You may also be able to find a career path not listed. Choose three possible career paths and investigate what you would need to do to be ready to fill one of those positions.

Among the items to look for:

- Education—college? What would you major in? Should you have a minor?

- Working conditions—What is the day to day job like? Will you have to be out in the field all day, or is it a desk job? Is the work strenuous? Are you working on a drilling rig at sea, away from your family for 3–6 months at a time? Is the job "hazardous"?

- Pay scale—What is the average "starting pay" for the position you are seeking? Don't be fooled by looking at "average pay," which may include those who have been working for 20+ years.

- Is the job located in the states, or is there a possibility to travel to other parts of the world?

- Are there any other special requirements?

Architect
Architectural Technician or Technologist
Botanist
Bricklayer
Building Surveyor
Building Technician
Carpenter or Joiner
Civil Engineer
Civil Engineering Technician
Construction Manager
Fence Installer
Firefighter
Fire Inspector/Investigator
Forest Fire Prevention Specialist
Fire Captain
Forest Ranger
Land Surveyor
Landscape Manager
Landscape Scientist
Landscaper
Roofer
Rural Surveyor
Scaffolder
Steel Erector
Steel Fixer
Steeplejack or Lightning Conductor Engineer
Stonemason
Structural Engineer
Thermal Insulation Engineer
Tiler
Town Planner
Town Planning Support Staff
Window Fitter
Wood Machinist

Team Building Activity

In a team of four students, create a before and after scale model of a house and property showing improvements to make the property resistant to an approaching wildfire. Two students will create the "before" model and two will create the "after" model. Both models should complement each other (houses should be nearly identical and the property footprint should also be nearly identical).

ANTICIPATION GUIDE FOR:
TECHNOLOGY: FULL TILT

Purpose: To identify what you already know about architecture design and restoration, to direct and personalize your reading, and to provide a record of what new information you learned.

Before you read about the Leaning Tower of Pisa in "Technology: Full Titlt," examine each statement below and indicate whether you agree or disagree. Be prepared to discuss your reactions to the statements in groups.

- In Pisa, Italy, there is a leaning tower.
- Construction of the 185 foot "Leaning Tower" began in 1173 and was completed in twenty years.
- The famous "Leaning Tower" was designed to tilt.
- The tower is tilting because the original foundation rests on soft sand and clay.
- The famous "Leaning Tower" has always leaned in the same direction (towards the south).
- Since the towers in Pisa tilt, all buildings will also be tilted.
- Attempts to stabilize the tower have been ongoing for two hundred years.
- In the past several years, engineers have drilled and removed some of the sand and clay below the foundation, allowing the tilt of the famous tower to move back towards the north.
- The famous "Leaning Tower" tilts more than sixteen feet off center.
- The government in Pisa wants to completely straighten the lean of the famous tower.
- With the success of stabilizing the famous "Leaning Tower," engineers are working on stabilizing the other towers in Pisa.

TECHNOLOGY:
FULL TILT

By Catherine L. Barker

ONLY TIME WILL TELL
IF HUMAN EFFORT TO SAVE THESE LANDMARKS WILL COME TO A TOPPLING END.

One hopes the Leaning Tower of Pisa won't someday be the only Tower of Pisa.

Of all Pisa's leaning towers—yes, there are several—the famous one is the least likely to topple. That's because an 11-year restoration effort, involving three years of painstaking soil removal, has successfully steadied the precariously poised campanile.

Pisa's soil is mostly compressible clay and sand, which gives way over time and causes big buildings to shift. The iconic edifice started listing northward during its first phase of construction, in the 1100s, then changed course, pitching southward over the next eight centuries. An 1817 measurement put its incline at 5 degrees; by 1990, the cant had increased to 5.5. Fearing the 197-foot-tall, tourist-luring monument might collapse, Italy's premier formed an international team to preserve it.

John Burland, a top project engineer, says the tower's tilt is back to 5 degrees, and "over the last two years, almost no movement has been detected." The city's other bell towers, though linked to larger structures, haven't been bolstered. One hopes the Leaning Tower of Pisa won't someday be the only Tower of Pisa.

Adapted from "Technology: Full Tilt" by Catherine Barker: National Geographic Magazine, August 2009.

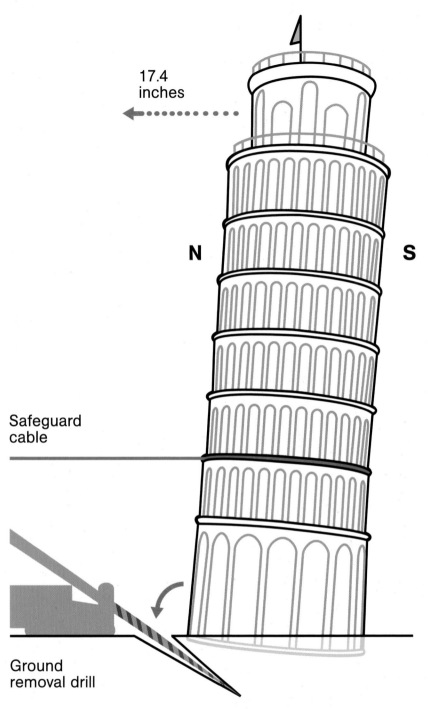

17.4 inches

N

S

Safeguard cable

Ground removal drill

STRAIGHT STORIES Drilling in 41 places helped level the base, which shifted the tower top nearly 1.5 feet.

Art: Mariel Furlong, NG Staff, NGM Maps

La Torre di Pisa
Year completed: circa 1370
Tilt: 5 degrees

© Gianluca Colla/National Geographic Stock

San Michele degli Scalzi
Year completed: circa 1100
Tilt: 5 degrees

San Nicola
Year completed: circa 1250
Tilt: 2.5 degrees

Writing Assignment

After reading the article "Technology: Full Tilt" and conducting your own research, write a brief paper discussing the design and construction of the famous Leaning Tower of Pisa.

In addition to your own ideas and thoughts, please discuss:

- The timeline for the construction of the famous Leaning Tower.

- The reason for construction delays.

- The original purpose of the tower.

- The remediation efforts to stop the progress of the leaning.

- Other than Pisa, Italy, are there other "leaning towers?"

Listed below are a few links to help you in getting started.

Links:

Tower of Pisa information
http://www.towerofpisa.info/

Travelogue of Italy
http://www.italylogue.com/planning-a-trip/leaning-tower-of-pisa.html

Frommer's Travel Guide of Pisa
http://www.frommers.com/destinations/pisa/0168010029.html

TLC – Pisa
http://tlc.howstuffworks.com/family/leaning-tower-of-pisa-landmark.htm

UK-Telegraph Solves the 800 Year Mystery
http://www.telegraph.co.uk/culture/art/architecture/7907298/Solving-the-800-year-mystery-of-Pisas-Leaning-Tower.html

Hydrogeology/Geochemistry of Pisa
http://www.sciencedirect.com/science/article/pii/S0883292704001611

Geology Correlations Pisa
http://www.aiqua.it/images/directory/130864183508SP04_Sarti.pdf

Geodynamic Process Pisa
http://www.dst.uniroma1.it/sciterra/sezioni/doglioni/Publ_download/SP409083-096%20Corti.pdf

Career Investigation

Imagine pouring yourself a tall glass of water and setting it on the table, only to notice that the water spills over the lip of the glass on one side. You measure the table and discover all the legs are exactly the same height. The problem appears to be that the floor is sloped by more than five degrees. How could that be? Your building was level when it was built, so what could have happened? Quite possibly your building has subsided.

Listed here are several possible careers to investigate. You may also be able to find a career path not listed. Choose three possible career paths and investigate what you would need to do to be ready to fill one of those positions.

Among the items to look for:

- Education—college? What would you major in? Should you have a minor?

- Working conditions—What is the day to day job like? Will you have to be out in the field all day, or is it a desk job? Is the work strenuous? Are you working on a drilling rig at sea, away from your family for 3–6 months at a time? Is the job "hazardous"?

- Pay scale—What is the average "starting pay" for the position you are seeking? Don't be fooled by looking at "average pay," which may include those who have been working for 20+ years.

- Is the job located in the states, or is there a possibility to travel to other parts of the world?

- Are there any other special requirements?

Architect
Building Surveyor
Building Technician
Carpenter or Joiner
Carpet Fitter/Floor Layer
Ceiling Fixer
Civil Engineer
Civil Engineering Technician
Concrete Supervisor
Concrete Finisher
Construction Manager

Geoscientist
Geotechnician
Land Surveyor
Landscape Manager
Landscape Scientist
Landscaper
Mechanical Engineering Technician
Planning and Development
Surveyor
Research Scientist
Structural Engineer
Technical Surveyor
Town Planner
Town Planning Support Staff

Team Building Activity

You will be assigned to a group of four students, where you will be a part of a panel discussing possible plans to remediate the famous leaning tower so that it becomes completely vertical. Each person on the panel will assume the role of geologist, structural engineer, city official, and tourism official. Students in the audience will assume the role of news reporters, interested townspeople, and affected tourism companies. The panel will make a brief presentation regarding the plans to remediate the famous Leaning Tower and then take questions from the public.

ANTICIPATION GUIDE FOR:
VILLAGE GREEN

Purpose: To identify what you already know about renewable and nonrenewable resources, to direct and personalize your reading, and to provide a record of what new information you learned.

———————◆———————

Before you read about carbon footprints and global energy use in "Village Green," examine each statement below and indicate whether you agree or disagree. Be prepared to discuss your reactions to the statements in groups.

- We can eliminate 'vampire power' by making sure that all electronics stay plugged into electrical outlets.
- Renewable energy is an energy source that can be reused.
- Clean, or green, energy sources include solar, wind, and coal.
- The method used to capture CO_2 and house it in a geological formation is practical and safe.
- The light-emitting diodes used in traffic signals save not only thousands of dollars of power costs but also reduces millions of pounds of carbon dioxide emissions each year.
- Carbon dioxide is a harmful gas and should be eliminated.
- Solar energy is one of the cleanest energies; it doesn't emit any greenhouse gases, and it should be the main source of power for everyone.
- Since the inception of the Kyoto Protocol, greenhouse gases (and specifically CO_2) have decreased substantially.
- Designing, building, and maintaining a "green" house is easy, so easy, in fact, that the majority of homes built today are near zero emission homes.
- Volcanic eruptions emit more greenhouse gases than all of the man-made emissions.
- The government should impose a tax or heavy fines for companies that emit greenhouse gases into the atmosphere.
- The greenhouse gas issue is too big for anyone to tackle; there is little that "I" can do.

VILLAGE
GREEN

By Michelle Nijhuis

Denmark's tiny island of Samsø is one of the few communities on Earth with almost no carbon footprint. In 1997 the Danish government challenged citizens to strive for energy self-sufficiency. Today 11 wind turbines provide electricity for the island's roughly 4,000 residents; 10 more produce power to sell to the mainland. Other communities are following suit in England, Sweden, and elsewhere

WHILE BIG, HIGH-TECH ITEMS LIKE SOLAR FARMS OR GIANT WIND TURBINES GET THE LION'S SHARE OF THE PRESS,

CUTTING GREEN HOUSE EMISSIONS WILL TAKE MYRIAD SOLUTIONS,

BIG AND SMALL.

My carbon footprint should be petite. I live in a house lit by solar panels and snugly insulated with straw bales. I shower and wash dishes with the help of a superefficient water heater. I eat eggs laid near my home in the Rocky Mountains. Yet it's hard to be smug, for as soon as I leave my doorstep, my footprint swells several sizes. I drive to the grocery store. I fly on planes. I buy my shoes and solar-fed appliances from faraway factories that run on coal. Energy-thrifty alternatives, when they exist, are prohibitively time-consuming or pricey. So while climate protection starts at home, finishing the job calls for changing the systems around us.

We can't all crowd onto Samsø, the tiny Danish island that uses the sun and wind to liberate itself from greenhouse gas emissions. And most U.S. commuters can only dream about a place like Freiburg, Germany, where excellent public transportation and bike-friendly streets make cars virtually superfluous. Not to mention the city's ubiquitous solar panels that have helped them cut their emissions 10 percent over the past decade.

Communities around the world are joining the effort to stabilize the climate.

Samsø and Freiburg are pioneers, but they're no longer alone. Communities around the world are joining the effort to stabilize the climate. In the United States some 780 cities have pledged to meet the Kyoto Protocol's greenhouse gas reduction targets, even though President George W. Bush refused to sign the international agreement. From Los Angeles to Shepherdstown, West Virginia, from Miami to Milan, Minnesota, mayors have promised to shrink their cities' emissions to 7 percent below 1990 levels by 2012. Together they represent almost one in four Americans.

For Seattle Mayor Greg Nickels, climate change became a pressing concern in 2005, when a nearly snowless winter stripped Pacific Northwest ski slopes and squeezed the region's water supplies. Nickels knew that a single city would be foolhardy to tackle a global problem alone. "If only Seattle made the commitment, it would be purely symbolic, and it's hard to ask people to change their lives for something

Adapted from "Village Green" by Village Green: National Geographic Magazine, April 2008.

symbolic," he says. So Nickels challenged fellow mayors to join him.

Then the hard work began: improving bus service and bike lanes, expanding the city's green-building program, and placing an unpopular tax on parking in downtown commercial lots. Such efforts, along with major purchases of wind energy by the city utility, helped Seattle's government cut its emissions by almost 60 percent below 1990 levels. The entire city cranked down its greenhouse gas output by 8 percent from 1990 figures, beating Kyoto targets despite robust growth.

Nickels held a conference in Seattle, where mayors from around the country touted their accomplishments and traded ideas for doing more. Some are turning methane gas from their landfills into electricity, others are urging citizens to buy produce from local farm stands, and some are even pulling police officers out of patrol cars and putting them on Segways, the high-tech electric scooters. And while the goals aren't binding—and not all cities will meet them—reaching the modest Kyoto Protocol targets can be relatively painless, even profitable. Simply switching the bulbs in all a city's traffic signals to new efficient light-emitting diodes (LEDs) can save hundreds of thousands of dollars in power costs and millions of pounds of carbon dioxide emissions each year. Purchases of wind and solar energy can create local construction and engineering jobs, as they have in solar-crazed Freiburg.

But scientists say that if we're serious about stabilizing the climate, we'll need even deeper emissions cuts. So Nickels and his fellow mayors are now setting their sights on reducing citywide emissions 80 percent below 1990 levels by 2050—and calling for the federal government to help them with reforms and innovations needed to reach beyond city limits.

The cities aren't alone. As of the beginning of 2008, 18 U.S. states, representing almost 50 percent of the population, have established their own greenhouse gas reduction targets. California and Florida have set some of the most ambitious goals, aiming, like the mayors, to reduce emissions 80 percent below 1990 levels by 2050. States in the West, Northeast, and Midwest have teamed up to form regional climate-protection alliances. More than half of all states plan to boost their use of energy from wind turbines, solar panels, and geothermal sources, with several committed to using at least 20 percent renewable energy by 2020. California, a longtime leader in air-quality protection, is going further by both requiring electric utilities to adopt efficiency measures and rewarding the companies that do so, placing the profit motives of the utilities in line with the environmental goals of the state. The Golden State is also battling the federal Environmental Protection Agency for the right to regulate greenhouse gas emissions from new cars and trucks—putting automakers on notice that the state considers carbon dioxide a pollutant just as noxious as any other exhaust fume.

This patchwork of local, state, and regional efforts has caught the attention of another large and influential community: U.S. businesses. After years of denial—and, in some cases, deliberate obfuscation and misrepresentation of climate science—many large and small American corporations are now taking climate change very seriously.

Some companies got the message from thawing Arctic tundra or shrinking Rocky Mountain snowpacks. Others noted the growing consumer interest in environmentally friendly products or the clean-energy leanings of investment funds and venture capitalists. Many see the work of cities and states as a harbinger of national greenhouse gas regulations and want to be prepared.

"There's a sense that market transformation is inevitable, because mandatory regulation of greenhouse gases is inevitable," says Truman Semans of the Pew Center on Global Climate Change, a Virginia–based environmental think tank.

Corporate climate action breeds its share of bombast. When Fiji Water announced its plans to go "carbon negative," critics said the U.S. company—which imports boutique bottled water from the South Pacific at considerable energy cost—was a dubious climate-protection leader. The British ad standards agency recently reprimanded Royal Dutch Shell for an advertisement claiming, "We use our waste CO_2 to grow flowers," after environmentalists pointed out that the company sent only a minuscule percentage of its carbon dioxide emissions to greenhouses.

Yet Shell and other large corporations, such as Hewlett-Packard and DuPont, have set and attained various emissions-reduction goals for their own company practices, getting rid of inefficiencies and realizing profits in the process. Other companies are scrutinizing their global supply chains. Wal-Mart, whose market influence—and carbon footprint—knows few rivals, has begun a program to reduce the energy used in making and transporting toothpaste, soap, and a handful of its other basic items.

Some businesses, such as Whirlpool, are building more energy-efficient products, while others are creating new carbon-thrifty technologies. DuPont and BP, for instance, are working to overcome technical barriers to biobutanol, a renewable fuel with a higher energy content than ethanol. Shell is betting that algae-based biofuel may power the future, and Google has announced plans to invest

More than 770 U.S. cities have pledged to meet the Kyoto Protocol's greenhouse gas reduction targets, covering one in four Americans.

hundreds of millions of dollars in the development of cheap, renewable energy from the sun, wind, and other sources.

Cities and corporations, despite their laudable efforts, continue to dump plenty of carbon dioxide and other greenhouse gases into the atmosphere, which will alter the climate for decades, if not centuries, to come. After all, this is still a world where Peabody Coal sells seven or eight tons of fossil carbon each second. Bigger changes, say many in the business community, won't happen unless the U.S. government puts a price on carbon, whether through a tax or a cap-and-trade system for greenhouse gas emissions. "Voluntary efforts alone will not solve the problem," DuPont CEO Chad Holliday told a Senate committee last year.

Just before the latest round of international climate talks in Bali, leaders of more than 150 multinational companies issued a communiqué calling for "strong, early action" on climate change, including ambitious, legally binding international emissions restrictions and an expanded carbon market. "We believe that tackling climate change is the pro-growth strategy," they wrote. "Ignoring it will ultimately undermine economic growth."

For corporate America, which once ferociously opposed any form of carbon regulation, such talk is revolutionary. Though Seattle Mayor Greg Nickels is delighted by the audacity of cities, states, and corporations, he knows their efforts won't suffice. "This is an issue that's going to require a sense of national purpose, a sense that we're all in it together," he says. In other words, leadership from the top. But until Congress—and the President—answers the clamor for change, action will have to keep trickling up from below.

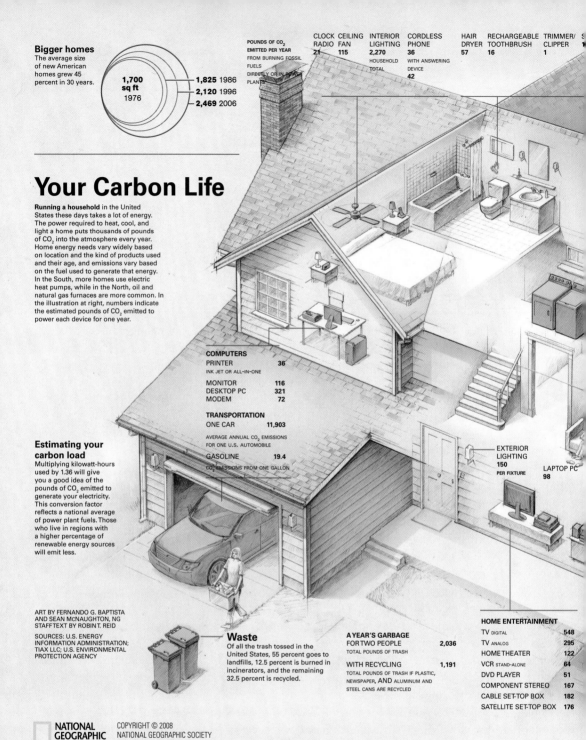

Bigger homes

The average size of new American homes grew 45 percent in 30 years.

1,700 sq ft 1976

- **1,825** 1986
- **2,120** 1996
- **2,469** 2006

Your Carbon Life

Running a household in the United States these days takes a lot of energy. The power required to heat, cool, and light a home puts thousands of pounds of CO_2 into the atmosphere every year. Home energy needs vary widely based on location and the kind of products used and their age, and emissions vary based on the fuel used to generate that energy. In the South, more homes use electric heat pumps, while in the North, oil and natural gas furnaces are more common. In the illustration at right, numbers indicate the estimated pounds of CO_2 emitted to power each device for one year.

Estimating your carbon load

Multiplying kilowatt-hours used by 1.36 will give you a good idea of the pounds of CO_2 emitted to generate your electricity. This conversion factor reflects a national average of power plant fuels. Those who live in regions with a higher percentage of renewable energy sources will emit less.

ART BY FERNANDO G. BAPTISTA AND SEAN MCNAUGHTON, NG STAFF TEXT BY ROBIN T. REID

SOURCES: U.S. ENERGY INFORMATION ADMINISTRATION; TIAX LLC; U.S. ENVIRONMENTAL PROTECTION AGENCY

POUNDS OF CO_2 EMITTED PER YEAR FROM BURNING FOSSIL FUELS DIRECTLY OR IN POWER PLANTS

CLOCK RADIO **21**	CEILING FAN **115**	INTERIOR LIGHTING **2,270**	CORDLESS PHONE **36** HOUSEHOLD TOTAL	HAIR DRYER **57**	RECHARGEABLE TOOTHBRUSH **16**	TRIMMER/ CLIPPER **1**

WITH ANSWERING DEVICE **42**

COMPUTERS

PRINTER	**36**

INK JET OR ALL-IN-ONE

MONITOR	**116**
DESKTOP PC	**321**
MODEM	**72**

TRANSPORTATION

ONE CAR	**11,903**

AVERAGE ANNUAL CO_2 EMISSIONS FOR ONE U.S. AUTOMOBILE

GASOLINE	**19.4**

CO_2 EMISSIONS FROM ONE GALLON

EXTERIOR LIGHTING 150 PER FIXTURE

LAPTOP PC **98**

Waste

Of all the trash tossed in the United States, 55 percent goes to landfills, 12.5 percent is burned in incinerators, and the remaining 32.5 percent is recycled.

A YEAR'S GARBAGE FOR TWO PEOPLE 2,036 TOTAL POUNDS OF TRASH

WITH RECYCLING 1,191 TOTAL POUNDS OF TRASH IF PLASTIC, NEWSPAPER, AND ALUMINUM AND STEEL CANS ARE RECYCLED

HOME ENTERTAINMENT

TV DIGITAL	**548**
TV ANALOG	**295**
HOME THEATER	**122**
VCR STAND-ALONE	**64**
DVD PLAYER	**51**
COMPONENT STEREO	**167**
CABLE SET-TOP BOX	**182**
SATELLITE SET-TOP BOX	**176**

84 ARCHITECTURE AND CONSTRUCTION

WASHER	DRYER	DRYER
153	ELECTRIC	GAS
	1,521	435

Vampire electronics

In nearly any home, you'll find electricity vampires—devices that suck power when you think they're off. They use standby power to respond to the push of a button. According to Lawrence Berkeley National Laboratory scientist Alan Meier, vampires consume 5 to 8 percent of a household's total energy use. To combat vampires, unplug them or use power strips with off switches.

LEADING CULPRITS
POUNDS OF CO_2 EMITTED PER YEAR FROM ELECTRONICS IN STANDBY MODE

GARAGE-DOOR OPENER	54	MONITOR, FLAT-SCREEN	14
MODEM, CABLE	46	MONITOR, CATHODE-RAY	10
MODEM, DSL	16	CORDLESS PHONE	12
MICROWAVE OVEN	37	PHONE CHARGER	3
DESKTOP PC	34		
INK-JET PRINTER	15		

FOR MORE INFORMATION ON STANDBY ELECTRONICS GO TO *STANDBY.LBL.GOV.*

KITCHEN	
EXHAUST FAN	20
MICROWAVE	179
RANGE ELECTRIC	628
RANGE GAS	655
TOASTER	53
TOASTER OVEN	45
DISHWASHER	599

DOES NOT INCLUDE HEATING WATER

COFFEEMAKER	83
REFRIGERATOR	1,191
FREEZER	1,397

Powering a Nation

Electricity is clean, but many of the fuels used to make it aren't—and our use of electric devices comes with a hidden environmental cost. Coal plants generate half the nation's electricity and account for a quarter of U.S. CO_2 emissions.

The electric home

The average amount of electricity used per U.S. home rose sixfold between 1950 and 2006, the last year for which data are available.

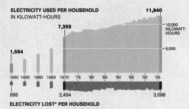

ELECTRICITY USED PER HOUSEHOLD
IN KILOWATT-HOURS

1,584   7,355   11,840

10,000 KILOWATT-HOURS
5,000

1950 1955 1960 1965 1970 '75 '80 '85 '90 '95 '00 '05

886   2,494   3,598

ELECTRICITY LOST* PER HOUSEHOLD

* ELECTRICITY LOST IN TRANSMISSION AND DISTRIBUTION FROM POWER PLANTS, OR OTHERWISE UNACCOUNTED FOR

A growing energy appetite

The total amount of energy used by American homes nearly tripled from 1950 to 2006, growing faster than the nation's overall energy use.

21	18	32	28

99 QUADRILLION BTU* USED IN 2006

6	16	8
4		

RESIDENTIAL
COMMERCIAL
INDUSTRIAL
TRANSPORTATION

1 QUADRILLION = 1 BILLION x 1 MILLION

34 IN 1950

* THE BRITISH THERMAL UNIT (BTU) IS USED TO MEASURE THE ENERGY CONTENT OF FUELS AND THE POWER OF HEATING AND COOLING SYSTEMS. ONE KILOWATT-HOUR OF ELECTRICITY IS EQUIVALENT TO 3,413 BTU. USING A 100-WATT LIGHT FOR ONE HOUR USES 341 BTU OF ENERGY.

HOME HEATING AND COOLING	
CENTRAL AC	4,067
ROOM AC	872
WATER HEATER ELEC.	3,586
WATER HEATER GAS	2,171
WATER HEATER OIL	4,331
FURNACE GAS	6,967
FURNACE OIL	14,380
HEAT PUMP ELECTRIC	5,249
FURNACE FAN	606

WITH OIL AND GAS FURNACES

Old king coal

Fossil fuels, including coal, supply 70 percent of U.S. energy needs. Power plants in Texas, Ohio, and Florida emit the most greenhouse gases, but sparsely populated, coal-rich states like Wyoming, North Dakota, and West Virginia top the per capita emissions list.

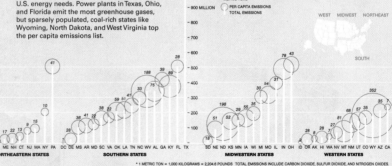

44

1 BILLION METRIC TONS
900 MILLION
800
700
600
500
400
300
200
100

2006 TOTAL POWER PLANT EMISSIONS
METRIC TONS*

PER CAPITA EMISSIONS
TOTAL EMISSIONS

WEST   MIDWEST   NORTHEAST

SOUTH

NORTHEASTERN STATES: VT RI ME NH CT NJ MA NY PA
SOUTHERN STATES: DC DE MS AR MD SC VA OK LA TN NC WV AL GA KY FL TX
MIDWESTERN STATES: SD NE ND KS MN IA WI MI MO IL IN OH
WESTERN STATES: ID OR AK HI WA NV MT NM UT CO WY AZ CA

* 1 METRIC TON = 1,000 KILOGRAMS = 2,204.6 POUNDS   TOTAL EMISSIONS INCLUDE CARBON DIOXIDE, SULFUR DIOXIDE, AND NITROGEN OXIDES.

7				3		
8	56	30		33		

99 QUADRILLION BTU* USED IN 2006   **35 IN 1950**

▓ RENEWABLE
▓ NUCLEAR
▓ FOSSIL FUELS
▓ IMPORTS

1 QUADRILLION = 1 BILLION x 1 MILLION

* THE BRITISH THERMAL UNIT (BTU) IS USED TO MEASURE THE ENERGY CONTENT OF FUELS AND THE POWER OF HEATING AND COOLING SYSTEMS. ONE KILOWATT-HOUR IS EQUIVALENT TO 3,413 BTU OF ELECTRICITY. USING A 100-WATT LIGHT FOR ONE HOUR USES 341 BTU OF ENERGY.

A Greener Way

Cutting household energy use is a win-win proposition, reducing your carbon footprint and the bite energy costs put on your wallet. Some fixes, like lowering your thermostat to 68°F or less in winter or raising it to 78°F or more in summer, offer immediate reductions. Others, such as replacing an old appliance with a more efficient model, cost more up front but rack up major savings over the life of the product. Often tax credits and incentives can help offset the cost of energy-efficient upgrades or hybrid vehicles. Many homeowners can now even choose a renewable energy source from their utility company.

To learn more
For more carbon-cutting tips, explore the Web links below as well as the resources on pages 87-8.

U.S. DOE OFFICE OF ENERGY EFFICIENCY AND RENEWABLE ENERGY
eere.energy.gov/consumer

ENERGY STAR PRODUCTS
energystar.gov

SOLAR ENERGY
eere.energy.gov/solar

WATER SENSE
epa.gov/watersense

HOME INSULATION
ornl.gov/sci/roofs+walls/insulation

CLEAN ENERGY
epa.gov/cleanenergy
epa.gov/greenpower/buygp/index.htm

SOLAR AND OTHER WATER HEATERS
eere.energy.gov/consumer/your_home/water_heating

WIND ENERGY
awea.org

TAX CREDITS AND INCENTIVES
dsireusa.org

PRESERVE OUR PLANET
nationalgeographic.com/preserveourplanet

NATIONAL GEOGRAPHIC GREEN GUIDE
thegreenguide.com

INSULATING GAS

MULTIPLE PANES

INSULATING FRAME MATERIALS

Windows
New insulated, double-pane models can slash energy loss by 40 percent per window.

Heating or cooling loss from the average single-pane window is responsible for about 250 pounds of CO_2 emissions a year.

Home wind power
In windy areas, new small-scale wind turbines can provide most of a home's energy needs for the year.

Eradicate le
Caulking leaks around window doors, and vent an easy way to energy losses.

Waste (water) not
Water treatment is a major energy drain for cities. A typical U.S. home uses 90,000 gallons a year. Newer fixtures improve on 1992 federal standards.

	SHOWER	FAUCET	TOILET
PRE-1992	3 GPM*	3 GPM	3.5 GPF*
1992	2.5 GPM	2.2 GPM	1.6 GPF
HIGH-EFFICIENCY	<2 GPM	1.5 GPM	1.28 GPF

* GALLONS PER MINUTE / GALLONS PER FLUSH

Protecting
Uninsulated d lose 10 to 30 p the energy use and cool the a carry.

BATTERY

HYBRID ENGINE

Sip, don't guzzle
Burning one gallon of gasoline releases 19.4 pounds of CO_2. Better fuel efficiency for the 250 million vehicles on U.S. roads could cut emissions significantly.

MPG	CO_2	ASSUMES 12,000 MILES DRIVEN PER YEAR
20	11,688	AVERAGE MPG FOR ALL U.S. CARS IN 2001
22.2	10,530	CURRENT U.S. STANDARD FOR LIGHT TRUCKS
27.5	8,500	CURRENT U.S. STANDARD FOR CARS
32	7,305	PROPOSED LIGHT TRUCK STANDARD
38.3	6,104	PROPOSED CAR STANDARD

Cycle
Bicycles were the main transportation to and from work for 705,000 Americans in 2005.

Home energy audit
Energy auditors can identify dozens of ways homes lose energy, from leaks to outdated insulation. Many utility companies provide audits or can recommend independent specialists.

▣ NATIONAL GEOGRAPHIC CHANNEL

NATIONAL GEOGRAPHIC
GREEN GUIDE

PRESERVE OUR
PLANET

Solar power

Photovoltaic panels are still relatively expensive, but their costs are falling and their efficiency is rapidly improving. Excess electricity can be sold back to power companies, and some state and federal tax breaks are available.

Lighter lighting

All those bulbs add up: After heating and cooling, lights are one of the biggest energy users in your home. Compact fluorescents can slash energy use, and highly efficient light-emitting diodes (LEDs), now seen in flashlights and car taillights, are just entering the residential-lighting market with the possibility of greater savings.

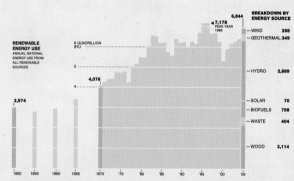

	POWER USED	LIFETIME	CO₂ EMISSIONS*
60-WATT INCANDESCENT	LIGHT 6 WATTS / HEAT 54	750 TO 1,000 HOURS	95.3 POUNDS
13-WATT COMPACT FLUORESCENT	5 / 8	6,000 TO 10,000	20.6

* IF USED 3 HOURS PER DAY FOR A YEAR

Warmer walls

Sprayed foam insulation can cut heat loss and plug gaps by completely filling wall spaces around plumbing and electrical conduits.

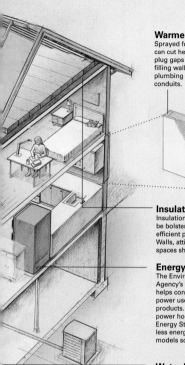

Insulation

Insulation in older homes often can be bolstered with the more energy-efficient products made today. Walls, attics, basements, and crawl spaces should be insulated.

Energy Star

The Environmental Protection Agency's Energy Star program helps consumers reduce their power use through energy-efficient products. Most refrigerators are power hogs, but those carrying the Energy Star label use 40 percent less energy than conventional models sold in 2001.

Water heaters

Heating water can burn up 10 percent or more of a home's energy bill. Setting thermostats to 120°F and installing insulating blankets can cut energy use, as can solar heaters and tankless on-demand systems.

Using Renewable Power

Though wood and hydroelectric power are still the largest sources of renewable energy, solar and wind power are the fastest growing. Wind power alone generated enough electricity to support 2.3 million typical American homes in 2005.

RENEWABLE ENERGY USE
ANNUAL NATIONAL ENERGY USE FROM ALL RENEWABLE SOURCES

2,974   4,076   7,178 PEAK YEAR 1996   6,844

BREAKDOWN BY ENERGY SOURCE
- WIND 258
- GEOTHERMAL 349
- HYDRO 2,889
- SOLAR 70
- BIOFUELS 758
- WASTE 404
- WOOD 2,114

1950 1955 1960 1965 1970 '75 '80 '85 '90 '95 '00 '05

Green generation

Nearly every state produces renewable energy, with nine getting at least 20 percent of their power from renewable sources. To find renewable-energy suppliers in your state, go to *epa.gov/greenpower/pubs/gplocator.htm.*

RENEWABLE ENERGY SOURCES
PERCENTAGE OF TOTAL ELECTRICITY GENERATED IN 2006
- EQUAL TO OR HIGHER THAN U.S. AVERAGE
- LOWER THAN U.S. AVERAGE

WA 79% • MT 39% • ND 7% • MN 5% • WI 6% • MI 4% • NH 15% • VT 26% • ME 46% • OR 79% • ID 94% • WY 3% • SD 60% • IA 4% • IN 0% • OH 1% • NY 21% • MA 7% • RI 2% • CT 6% • NJ 2% • DE 0% • MD 5% • DC 0% • NV 12% • UT 2% • CO 4% • NE 5% • KS 0% • MO 2% • KY 4% • WV 1% • VA 5% • PA 2% • NC 6% • CA 31% • AZ 10% • NM 1% • OK 5% • AR 10% • TN 11% • SC 5% • AL 11% • GA 6% • MS 4% • LA 4% • TX 1% • FL 3% • AK 21% • HI 8%

Writing Assignment

After reading the article "Village Green" and conducting your own research, write a brief paper discussing the negative consequences of greenhouse gases in our atmosphere.

In addition to your own ideas and thoughts, please discuss:

- What is meant by "carbon footprint"?

- What is carbon and why is carbon important in our lives?

- What is carbon dioxide (CO_2) and what are the benefits and detriments of carbon dioxide?

- How has the level of carbon dioxide changed in the past one hundred years?

Career Investigation

As the world struggles to reign in our ever growing greenhouse gas emissions, there will be a huge demand for a skilled workforce.

Listed here are several possible careers to investigate. You may also be able to find a career path not listed. Choose three possible career paths and investigate what you would need to do to be ready to fill one of those positions.

Among the items to look for:

- Education—college? What would you major in? Should you have a minor?

- Working conditions—What is the day to day job like? Will you have to be out in the field all day, or is it a desk job? Is the work strenuous? Are you working on a drilling rig at sea, away from your family for 3–6 months at a time? Is the job "hazardous"?

- Pay scale—What is the average "starting pay" for the position you are seeking? Don't be fooled by looking at "average pay," which may include those who have been working for 20+ years.

- Is the job located in the states, or is there a possibility to travel to other parts of the world?

- Are there any other special requirements?

Biofuel Technology Development Manager
Biologist
Cartographers
Chemical Engineer
Civil Engineer
Civil Engineering Technician
Climate Change Analyst
Deep Sea Diver
Emergency Management Specialist

Environmental Compliance Inspector
Environmental Engineer
Environmental Engineering Technician
Environmental Scientist
Fuel Cell Engineer
Hydroelectric Plant Technician
Industrial Engineer
Geographer
Geoscientist
Hydrologist
Marine Biologist
Marine Engineer
Mechanical Engineer
Mechanical Engineering Technician
Methane Gas Generation System Technician
Nuclear Engineer
Park Ranger
Petroleum Engineer
Photonics Engineer
Photonics Technician
Solar Energy System Engineer

Solar Energy System Technician
Soil Scientist
Solid Waste Engineer
Sustainability Specialist
Water & Liquid Waste Treatment Engineer
Wind Energy Engineer
Wind Turbine Service Technician

Team Building Activity

You will be assigned to a group of four to five students, and will conduct an energy audit of a classroom or school building (as assigned by your instructor). You may use the diagrams in this article as a guideline. You may require assistance from building maintenance personnel or administration.

Upon completion of your energy audit, your team will make a presentation to school administrators on methods to improve the building's efficiency.

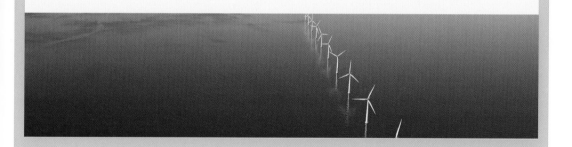